Religious Tourism and the Environment

CABI Religious Tourism and Pilgrimage Series

General Editors:

Dr Razaq Raj, Leeds Business School, Leeds Beckett University, UK
Dr Kevin Griffin, School of Hospitality Management and Tourism, Dublin Institute of Technology, Ireland

This series examines the practical applications, models and illustrations of religious tourism and pilgrimage management from a variety of international perspectives. Pilgrimage is not only a widespread and important practice in Islam, Judaism and Christianity, but also in other major religious traditions such as Buddhism, Hinduism and Sikhism.

The series explores the emergence and trajectories of religious tourism and pilgrimage. Inclusive of all denominations, religions, faiths and spiritual practices, it covers evaluations of religious tourism and pilgrimage, management guides, economic reports and sets of represented actions and behaviours within various cultural, management and marketing contexts. A key strength of the series is the presentation of current and diverse empirical research insights on aspects of religious tourism and pilgrimage, juxtaposing this with state-of-the-art reflections on the emerging theoretical foundations of the subject matter.

The series illustrates the principles related to religion, pilgrimage and the management of tourist sites. It aims to provide a useful resource for researchers and students of the subject, and increase understanding of this vital aspect of tourism studies.

Titles Available

Pilgrimage and Tourism to Holy Cities: Ideological and Management Perspectives
Edited by Maria Leppäkari and Kevin Griffin

Conflicts, Religion and Culture in Tourism
Edited by Razaq Raj and Kevin Griffin

The Many Voices of Pilgrimage and Reconciliation
Edited by Ian S. McIntosh and Lesley D. Harman

Local Identities and Transnational Cults within Europe
Edited by Fiorella Giacalone and Kevin Griffin

Religious Pilgrimage Routes and Trails: Sustainable Development and Management
Edited by Anna Trono and Daniel Olsen

Risk and Safety Challenges for Religious Tourism and Events
Edited by Maximiliano Korstanje, Kevin Griffin and Razaq Raj

Pilgrimage in Practice: Narration, Reclamation and Healing
Edited by Ian S. McIntosh, E. Moore Quinn and Vivienne Keely

Religious Tourism in Asia
Edited by Shin Yasuda, Razaq Raj and Kevin Griffin

Islamic Tourism: Management of Travel Destinations
Edited by Ahmad Jamal, Kevin Griffin and Razaq Raj

Managing Religious Tourism
Edited by Maureen Griffiths and Peter Wiltshier

Tourism, Pilgrimage and Intercultural Dialogue
Edited by Dolors Vidal-Casellas, Silvia Aulet and Neus Crous-Costa

Spiritual and Religious Tourism: Motivations and Management
Edited by Ruth Dowson, Jabar Yaqub and Razaq Raj

Dark Tourism and Pilgrimage
Edited by Daniel Olsen and Maximiliano Korstanje

Religious Tourism and the Environment
Edited by Kiran Shinde and Daniel Olsen

Religious Tourism and the Environment

Edited by

Kiran A. Shinde
La Trobe University

and

Daniel H. Olsen
Brigham Young University

CABI

CABI is a trading name of CAB International

CABI
Nosworthy Way
Wallingford
Oxfordshire OX10 8DE
UK

Tel: +44 (0)1491 832111
Fax: +44 (0)1491 833508
E-mail: info@cabi.org
Website: www.cabi.org

CABI
WeWork
One Lincoln St
24th Floor
Boston, MA 02111
USA
Tel: +1 (617)682-9015
E-mail: cabi-nao@cabi.org

References to Internet websites (URLs) were accurate at the time of writing.

A catalogue record for this book is available from the British Library, London, UK.

Library of Congress Cataloging-in-Publication Data
Names: Shinde, Kiran A., editor. | Olsen, Daniel H., 1973- editor.
Title: Religious tourism and the environment / edited by Kiran A. Shinde, Daniel H. Olsen.
Description: Wallingford, Oxfordshire ; Boston, MA : CAB International, 2020. | Includes bibliographical references and index. | Summary: "This is the first book to discuss both religious tourism and the environment. Moving beyond the documentation of impacts, it examines intersections between the two areas, such as the environment as a resource or generator of religious tourism. Chapters discuss such important topics as disease, environmental responsibility and host perspectives"-- Provided by publisher.
Identifiers: LCCN 2020023629 (print) | LCCN 2020023630 (ebook) | ISBN 9781789241600 (hardback) | ISBN 9781789241624 (epub) | ISBN 9781789241617 (ebook)
Subjects: LCSH: Tourism--Religious aspects. | Pilgrims and pilgrimages. | Tourism--Environmental aspects.
Classification: LCC G156.5.R44 R457 2020 (print) | LCC G156.5.R44 (ebook) | DDC 338.4/791--dc23
LC record available at https://lccn.loc.gov/2020023629
LC ebook record available at https://lccn.loc.gov/2020023630

ISBN-13: 9781789241600 (hardback)
9781789241617 (ePDF)
9781789241624 (ePub)

Commissioning Editor: Claire Parfitt
Editorial Assistant: Lauren Davies
Production Editor: Tim Kapp

Typeset by Exeter Premedia Services Pvt Ltd, Chennai, India
Printed and bound in the UK by CPI Group (UK) Ltd, Croydon, CR0 4YY

Contents

Contributors

Elisa Ascione is the Coordinator of the Center for Food and Sustainability Studies at the Umbra Institute, a US study abroad program located in the city of Perugia, Italy. Elisa teaches courses on Sustainability and Food Production, the Anthropology of Food and Eating, and History and Culture of Food in Italy. She has received an MA in Refugee Studies from the University of East London, UK, and a PhD in Anthropology from The University of Perugia. She has conducted research and published on heritagization processes of foods in Central Italy, and on the intersection of migration, work and gender relations in Italy. Center for Food and Sustainability Studies, The Umbra Institute, Piazza IV Novembre, 06123, Perugia, Italy. E-mail: eascione@umbra.org.

Robin A.E. Coningham holds UNESCO's Chair in Archaeological Ethics and Practice in Cultural Heritage at Durham University, UK, and has extensive experience of archaeology and post-disaster heritage interventions across southern Asia. He is interested in community engagement with archaeological excavations and site preservation, and the balance between heritage protection, pilgrimage and development. Department of Archaeology, Durham University, Durham, UK, DL1 3LE. E-mail: r.a.e.coningham@durham.ac.uk.

Michael A. Di Giovine is Associate Professor of Anthropology at West Chester University and Director of its Museum of Anthropology and Archaeology. He is also Honorary Fellow in the Department of Anthropology at the University of Wisconsin-Madison. A former tour operator, Michael is Convenor of the Anthropology of Tourism Interest Group at the American Anthropological Association. He has spoken and published widely on pilgrimage, tourism, and heritage; among his many publications are *The Seductions of Pilgrimage: Sacred Journeys Afar and Astray in the Western Religious Tradition* (Ashgate, 2015) and *Pilgrimage Beyond the Officially Sacred* (Routledge 2020). He is the Book Reviews Editor of *Journeys* and *The Journal of Tourism and Cultural Change*, and is editor of the book series *The Anthropology of Tourism: Heritage, Mobility and Society* (Lexington Books). West Chester University of Pennsylvania, 775 S. Church Street, West Chester, Pennsylvania, 19383, USA. E-mail: michael@michaeldigiovine.com.

Nour Farra-Haddad is a Senior Researcher and lecturer at St-Joseph University of Beirut, at The Lebanese University, and at the American University of Science and Technology in Lebanon. She is a specialist in pilgrimages and religious tourism. She founded and managed Lebanon Roots, the travel agency of the Lebanese Maronite Order, for five years, and currently manages her own travel consultancy company called NEOS. She holds a one PhD in Religious Anthropology, a Masters Degree in Anthropology, and two Bachelor Degrees in Archaeology and Sociology. She also holds a Diploma in Tour Guiding from the Ministry of Tourism, and is a founding member of the Association for the Development of Pilgrimages and Religious Tourism (APL). She is the author of two touristic guides: *Wiz Kids* and *Eco Lebanon: Nature and Rural Tourism*, and recently developed an app promoting Interreligious Tourism in Lebanon called "Holy Lebanon." Lecturer (USJ-FSR, UL, AUST), Beirut-Lebanon, Sin el Fil, Cityrama, Pasteur street, Vanlian Bldg No 12, 5th Floor, Lebanon. E-mail: nour@neoslb.com

Josep-Maria Garcia-Fuentes is an architect and Senior Lecturer in Architecture at the School of Architecture, Planning and Landscape of Newcastle University, UK. He is also a Fellow at the Catalan Observatory at the London School of Economics and a Visiting Professor at the Politecnico di Milano in Italy. Josep-Maria's PhD analysed the construction of modern Montserrat. He currently does research on architecture, architectural history, heritage, tourism and experimental preservation, and publishes and speaks widely about these topics and his own professional work at invited keynote lectures. School of Architecture, Planning and Landscape, Newcastle University, Newcastle upon Tyne, UK, NE1 7RU. E-mail: josep.garciafuentes@newcastle.ac.uk

Anouk Lafortune-Bernard has just completed a PhD in the Department of Archaeology at Durham University. She is interested in the social and economic impact of cultural heritage, the balance between heritage conservation and development, and community engagement. Her PhD thesis analysed the development of Lumbini, Birthplace of the Buddha and UNESCO World Heritage Site and its impact on surrounding local communities. She has also been involved in several research projects led by Durham University's UNESCO Chair, primarily at archaeological and World Heritage sites in Nepal, Sri Lanka, and India. Department of Archaeology, Durham University, Durham, UK, DL1 3LE, E-mail: a.lafortune91@gmail.com

Po-Hsin Lai is Senior Lecturer in Newcastle Business School at the University of Newcastle, Australia. Her research focuses on the dynamic processes of place meanings and landscape representations and implications for relevant stakeholders in the context of change driven by resource use activities such as tourism, outdoor recreation, mining, and urban sprawl, primarily in rural environments and protected areas. University of Newcastle, Newcastle Business School, University Drive, Callaghan NSW 2308, Australia. E-mail: po-hsin.lai@newcastle.edu.au.

Yang Mu is a recent graduate (MES in Tourism Policy and Planning) of the University of Waterloo, Canada. Her research interest is on geography of sacred landscapes, with particular emphasis on local meanings and interpretations of cultural and spiritual landmarks, and their significance for tourism development in the Everest region of Nepal. E-mail: SUNNY_MU1106@hotmail.com

Joshua Nash is an islophilic generalist-cum-linguist working on the language of Pitcairn Island. He writes about ethnography, the anthropology of religion, architecture, pilgrimage studies, and language documentation. He has conducted linguistic fieldwork on Pitcairn Island and Norfolk Island, South Pacific, Kangaroo Island, South Australia, and in New Zealand; environmental and ethnographic fieldwork in Vrindavan, India; and architectural research in outback Australia. He was Associate Professor at Aarhus Institute for Advanced Studies, Denmark in 2018–2019. Associate Professor, Aarhus Institute of Advanced Studies (AIAS), Aarhus University, 8000 Aarhus, Denmark, and Post-doctoral Research Fellow, University of New England, Armidale, NSW 2351, Australia. E-mail: jnash7@une.edu.au.

Sanjay K. Nepal is Professor in the Department of Geography and Environmental Management, University of Waterloo, Canada. His specialty is in tourism and conservation geographies, primarily working in remote and peripheral locations including high mountains in the Nepalese Himalaya and the Canadian Rockies. University of Waterloo, 200 University Ave W, Waterloo, Ontario, N2L 3G1, Canada. E-mail: snepal@uwaterloo.ca.

Daniel H. Olsen is an Associate Professor in the Department of Geography at Brigham Young University in Provo, Utah, USA. His research interests revolve around religious and spiritual tourism, heritage tourism, and the management of sacred sites, with secondary research interests in tourism in peripheral areas, and tourism and disabilities. He is co-editor of *Religion, Tourism, and Spiritual Journeys* (Routledge, 2006), *Religious Pilgrimage Routes and Trails* (CABI, 2018), and *Dark Tourism and Pilgrimage* (CABI, 2020), and has published over 50 journal articles and book chapters on religious and spiritual tourism. Department of Geography, 6–22 SWKT, Brigham Young University, Provo, Utah, USA, 84660. E-mail: dholsen@byu.edu.

Jahanzeeb Qurashi is a lecturer in Hospitality and Tourism at the University of Wales Trinity of Saint David, UK. He recently received a PhD in Tourism at the University of Central Lancashire, UK. He obtained his BA (Hons) degree from the University of Birmingham in Tourism and Business Administration, his first Masters degree in International Hospitality from the University of Birmingham, and a second Masters Degree (MBA) from University of Chester in Business and Marketing. He also holds a Postgraduate Certificate in research methodology from the University of Central Lancashire, UK. He is also the author of several articles and book chapters. His interests include commodification of religious tourism, the experiences of pilgrims and tourists, destination management, contemporary hospitality, event management, the role of SMART media technologies in the tourism and hospitality sector, digital marketing of tourism, and heritage tourism. He is a member of the editorial board for the *International Journal of Religious Tourism and Pilgrimage* and a member of the Institute for Religious Tourism and Pilgrimage. University of Wales Trinity of Saint David, Birmingham, UK. E-mail: j.qurashi@uwtsd.ac.uk.

Kiran Shinde teaches in the Planning program at La Trobe University, Australia. He does research in the field of religious tourism, and has published about 50 research articles relating to cultural heritage, religious tourism, policy analysis, and destination planning and management. His papers have appeared in high-ranking journals, including *Urban Studies, Current Issues in Tourism, International Journal of Tourism Research, GeoForum*, and *Space and Culture*. He has also contributed chapters in books published by Elsevier, Routledge, SUNY Press and CABI. He has presented research papers at conferences in Canada, Turkey, Australia, Portugal, Singapore, the UK, and the USA. He was invited by the UNWTO to contribute to the *Bethlehem Charter on Religious Tourism*. He is also on the editorial board of the *International Journal of Religious Tourism and Pilgrimage*. Recently, he completed a UNWTO report on Buddhist tourism in Asia. He holds a PhD from Monash University, Australia, an MSc from the Asian Institute of Technology, Thailand, and an M.Tech from CEPT University, India. College of Arts, Social Sciences and Commerce, La Trobe University, Australia. E-mail: k.shinde@latrobe.edu.au.

Rajendra N. Suwal is an Ashoka Fellow and has extensive experience on the biodiversity of Nepal – especially on bird ecology, Sarus Crane Conservation, wetland development and ecotourism. He is interested in capacitating citizen scientists, promoting community conservation stewardship, community-based wildlife tourism, and designing sustainable finance mechanisms in protected areas, strengthening national economy and benefitting local livelihood. Deputy Director, Partnership Development, WWF Nepal, Kathmandu, 44600. E-mail: Rajendra.suwal@wwfnepal.org; suwalraj@gmail.com.

Kai Weise has been working along the Himalayas as an architect and planner over the past three decades. He has facilitated the establishment of management systems for World Heritage properties in Nepal, Uzbekistan, India, and Myanmar, including for Lumbini. He has been involved in the strategic response and rehabilitation planning after the 2015 Gorkha Earthquake in Nepal and the 2016 Chauk Earthquake in Myanmar. President, ICOMOS Nepal, Kathmandu, E-mail: kai.weise@yahoo.com.

1 The Environmental Impacts of Religious Tourism

Kiran A. Shinde[1]* and Daniel H. Olsen[2]
[1]College of Arts, Social Sciences and Commerce, La Trobe University, Australia; [2]Brigham Young University

Introduction

Like other tourism niche markets, the religious tourism market has increased over the past two decades, with sacred sites in all faith traditions receiving record numbers of visitors (Olsen, 2013a; Butler and Suntikul, 2018). Some scholars and tourism experts suggest that, each year, between 200–600 million people travel for religious purposes (Timothy, 2011; World Tourism Organization, 2011), which has led scholars to investigate the linkages between religious/spiritual tourism and pilgrimage to sacred sites in different religions including:

- Islam (e.g. Din, 1989; Timothy and Iverson, 2006; Henderson, 2009, 2011; El Hanandeh, 2013; Bouyahya, 2016)
- Catholicism (e.g. Collins-Kreiner and Kliot, 2000; Fleischer, 2000; Voye, 2002; Pohoață *et al.*, 2013; Rodrigues and McIntosh, 2014; Rodríguez *et al.*, 2018)
- Protestantism (e.g. Collins-Kreiner and Kliot, 2000; Fleischer, 2000; Feldman, 2007; Ron and Feldman, 2009)
- Hinduism (e.g. Singh, 1997, 2005; Ghosal and Maity, 2010; Shinde, 2017, 2018; Pinkney and Whalen-Bridge, 2018)
- Buddhism (e.g. Hall, 2006; Goldberg, 2013; YES Bank, 2014; Geary and Mason, 2016; Chen *et al.*, 2017; Wong, 2018)
- Sikhism (e.g. Jutla, 2002, 2006)
- Judaism (e.g. Kosansky, 2002; Cohen Ioannides and Ioannides, 2006; Collins-Kreiner, 2010a; Collins-Kreiner and Luz, 2018)
- Shintoism (e.g. Wong *et al.*, 2013, 2016; Wong and Ryan, 2013; Nakanishi, 2018)
- The Bahai (e.g. Gatrell and Collins-Kreiner, 2006; Kreiner *et al.*, 2015)
- The Church of Jesus Christ of Latter-day Saints (e.g. Olsen, 2006b, 2013b, 2016, 2019b; Schott, 2010; Olsen and Timothy, 2018).

While a large body of work examines the interrelationships between pilgrimage, religion, and tourism from anthropological, theological, ethnographic, historical, geographical, and religious studies viewpoints (e.g. Timothy and Olsen, 2006; Badone and Roseman, 2004; Collins-Kreiner, 2010b; Stausberg, 2012), this is a young and growing field of academic inquiry.

One of the research areas that is sorely lacking is the relationships between religion, tourism, and the environment. With so many people travelling to religious sites and events for religious, educational, and leisure purposes,

*Corresponding author: k.shinde@latrobe.edu.au

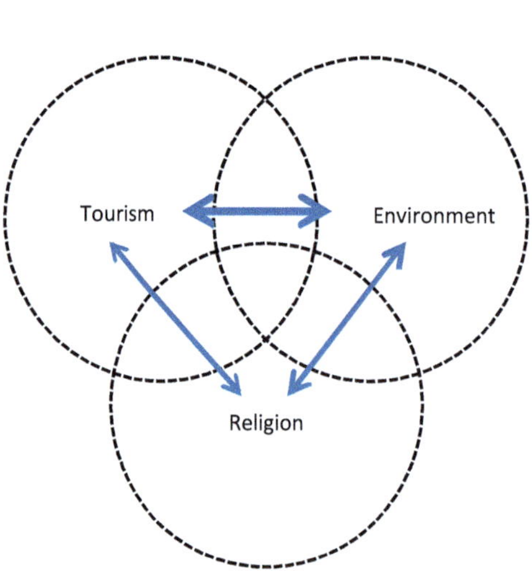

Fig. 1.1. The interrelationships between religion, tourism, and the environment.

what do these large movements and flows of visitors mean for sacred places of great religious importance? How do such large numbers of visitors interact with both the natural and the human-built environments of these destinations? How can natural and human-built environments support pilgrimage and religious tourism at such scales, and if they cannot, what can be done to increase the carrying capacities at these sites?

The purpose of this book is to examine the interrelationships between religion, tourism, and the environment (Fig. 1.1). This chapter sets the context for the other eleven chapters in this book by first briefly reviewing the literature between religion, tourism, and the environment, identifying gaps in the existing literature, and then presenting a conceptual model to understand the ecosystem of pilgrimage and religious tourism.

Tourism and Religion

There is a growing literature related to how religion and tourism influence each other (Timothy and Olsen, 2006). More particularly, this literature focuses on the similarities and differences between pilgrimage and religious tourism (Jackowski and Smith, 1992; Smith, 1992; Oberdick, 1995; Butler and Suntikul, 2018), with attempts to distinguish between leisure-inspired tourism and religiously motivated pilgrimages (Huntsinger and Fernández-Giménez, 2000; Bremer, 2000; Doron, 2005; Olsen, 2010; Shinde, 2012b; Damari and Mansfeld, 2016). These attempts have relied on the binary pairs of pilgrimage/tourism and pilgrim/tourist (Collins-Kreiner, 2010b; Olsen, 2010) binaries have also been used to illustrate the types of experiences different visitors have at sacred and religious sites (Fleischer, 2000; Poria *et al.*, 2003).

Many scholars have posited that religious tourism has its origins in religious practice of pilgrimage (Tomasi, 2002; Timothy and Olsen, 2006; Shinde, 2007b; Butler and Suntikul, 2018). Driven by religious needs and with religious and sacred places as its destination, pilgrimage is generally considered a form of tourism because of its use of existing tourism infrastructure for mobility purposes (Gupta, 1999; Timothy and Olsen, 2006). As Olsen and Timothy (2006, p. 7) argue, 'A "pilgrim" is a tourist (religious tourist) who is motivated by spiritual or religious factors'. Indeed, religious tourism combines elements of both religion and tourism, which combinations vary depending on a person's beliefs, needs, motivations, behaviour, and sought-after outcomes. However, there seems to be limited understanding of religious tourists– they are not a homogenous group (Olsen, 2013a). On the contrary, most religious sites will have a considerable blending of different types of religious tourists – from pilgrims who perform staunch rituals to those interested in the educational aspects of religious heritage to free-spirited travellers seeking to fulfill life goals through engaging with religious and spiritual practices (Huntsinger and Fernández-Giménez, 2000; Terzidou *et al.*, 2008; Moufakkir and Selmi, 2018; Rodríguez *et al.*, 2018). The nature of religious tourism and religious tourists differs based on religions, faiths, followers, religiosity, and the religious relevance of sites. The authors wonder if these different types of visitors lead to different kinds of impacts at religious tourism destinations. Moreover, do, different religious motivations translate into types of religious behaviour that have any bearing on the environmental impacts of religious tourism?

In recent years, the seasonality and visitation patterns within the religious tourism niche market have changed (Shinde, 2007a,

2017). For example, while religious rituals and performances are at the core of the pilgrimage economy that revolves around 'the concept of providing for the God and a sacred routine focusing on conspicuous consumption' (Rösel, 1983, p. 51), recreational activities are generally more hedonistic in nature. This mix of sacred and secular motivations and activities – a distinguishing feature of religious tourism – can have significant impacts on religious tourism destinations in terms of numbers of visitors, when these visitors come, and the activities in which they seek to participate. Also, outside of special religious events and feast days on religious calendars, most religious tourism destinations, like any other mass tourist destination, experience peak visitation during weekends, particularly by local and regional residents (Shinde, 2018). At the same time, as noted below, increasing religious tourism by international visitors is beginning to lead to a lack of seasonality, with religious tourism taking place throughout the entire week.

Some scholarship has focused on the broader philosophical aspects of these relationships, including cosmology, mythology, rituals, geography of sacred places, cultural performances, and religious-cultural heritage as related to religious tourism (e.g. McIntosh and Prentice, 1999; Olsen, 2003; Andriotis, 2011). However, until comparatively recently, little has been written on the more mundane aspects of religious tourism, such as infrastructure development, site management, and policy-making (e.g. Shackley, 2001; Petrillo, 2003; Raj and Morpeth, 2007; Rotherham, 2007; Shani *et al.*, 2007; Henderson, 2011; Shinde, 2012a). Also, even though religious tourism causes negative environmental impacts at host destinations (Holden, 2003; Olsen and Timothy, 2006; Shinde, 2007a), little has been written on these impacts of religious tourism both at local and broader geographical scales.

Tourism and Environment

At a basic level, pilgrimage refers to the movement of people to a religious or sacred site. In many cases, such as the Kumbha Mela in India, which has approximately 120 million participants (both pilgrims and tourists), the sheer scale and magnitude of these pilgrimages makes them no different from mass tourism in terms of their environmental impacts (Shinde, 2018).

There is a vast literature on the relationships between tourism and the environment (e.g. Holden and Fennell, 2012; Ballantyne and Packer, 2013; Mostafanezhad *et al.*, 2016; Dowling and Newsome, 2018). Within this literature, scholars have proposed several models to understand these relationships better. For example, Cohen (1978, p. 228) has presented a conceptual model suggesting four key ways tourism and the environment are related. These four ways include *the intensity of tourist site-use* (i.e. the numbers of visitors, duration of stay; their activities and facilities at their disposal); *the resiliency of the ecosystem* (i.e. the capability of the place to absorb the impacts); *the pace of development* (i.e. of tourist infrastructure); and *the transformational character of touristic developments* (i.e. attractions and further development). Cohen argued further that a combination of these factors would result in 'the accumulated environmental effects of growth, urbanization, commercialization, and functional diversification on the original tourist core areas'. Several studies have since illustrated the impacts on the natural and human-built sociocultural and physical environments of host destinations (e.g. Singh, 2002; Ambrósio, 2003; Alipour *et al.*, 2017).

While being debated, criticised, and modified over the past few decades (see Butler, 2006), another significant model is Butler's (1980) Tourist Area Life Cycle, which continues to provide a fundamental understanding of how increases in tourism flows alter the physical environment in a tourist destination. Based on carrying capacity concerns, this model outlines the different stages through which tourism development occurs. While this model has been widely used in leisure-oriented tourism to help develop appropriate tourism developmental and marketing polices, its application to religious sites has been fairly limited, with the exception of a comparative study of four religious destinations in Europe (Ambrósio, 2003) and a recent study of religious tourism in Mashdad, Iran (Alipour *et al.*, 2017).

Scholars have also used a systems approach to understand the relationships between tourism and environment and their growing complexities (Leiper, 1990; Holden, 2007; Buckley, 2011). While host destination environments attract tourists, the needs of these tourists and the tourism infrastructure developed to meet those needs alter these same environments. One model, proposed by Holden (2007, p. 9), notes that the tourism system has a range of different inputs, including nature and human resources, which resources are made available to consumers through a market system and regulated by government policies to attract investment. These inputs are combined with three distinct but interrelated subsystems – tourism retailing (e.g. corporate and independent travel agents and agencies), destinations (e.g. natural and cultural attractions, local transportation and accommodation infrastructure), and transportation (e.g. global airlines, car rental, and bus companies). These inputs and subsystems, when combined with different societal influence (e.g. changing consumer tastes, demographics, environmental and media, and technology), create a series of positive and negative outputs or outcomes that have the potential to either bring positive or negative cultural and environmental change to a destination. Holden's model also places a focus on broader contemporary environmental issues such as carbon emissions from flights, and tourist satisfaction.

Another stream of research regarding tourism—environment relationships is how tourists interact with and experience the environment. Several studies have shown that different types of tourists desire different environmental interactions as a part of their tourism experience (e.g. Young, 1999; Shoval, 2000; Andriotis, 2009). Iso-Ahola (1980), for example, identifies four major types of experiences tourists have with the environment in a leisure—recreational setting:

- environment as a setting for action (i.e. where interaction between tourists and the environment takes place)
- environment as a social system (i.e. where environment is the setting where social interaction and bonding takes place)
- environment as emotional territory (i.e. the emotional attachment people have to an environment)
- environment as self (i.e. where the environment and oneself become inseparable).

Research has also focused on the perceptions of residents regarding environmental tourism impacts (Liu *et al.*, 1987; Mesch and Manor, 1998; Jutla, 2000; Terzidou *et al.*, 2008). These studies reinforce the idea that while residents acknowledge the economic benefits of tourism, they are not always aware of the negative sociocultural and environmental impacts of tourism and their role in exacerbating these impacts (Dasgupta *et al.*, 2006; Gursoy *et al.*, 2004; Kamarudin and Nizam, 2013; Pohoaţă *et al.*, 2013).

Within this thematic literature, however, very little has been written regarding how religion intersects with tourism and the environment (see Holden, 2003; Olsen, Chapter 2, this volume; Timothy, 2012), although some research has focused on spiritual experiences in leisure outdoor settings (e.g.,Heintzman, 2009, 2014).

Religion and the Environment

In their edited book entitled *Religion and Environment*, Tanner and Mitchell (2002, p. 53) provide a working definition of religion 'as belief in a supernatural power or powers to be obeyed and worshipped and its expression in conduct and ritual'. They also define the environment as including 'all the natural features of land, water, flora and fauna which supports human life and influence its development and character'. Tanner and Mitchell note that 'the intensity of one's experience of their own environment depends to a large extent on its relevance to their livelihood and [the] religious beliefs with which they approach it'. These religious beliefs, along with their associated religious activities, have both direct/indirect and spatial/temporal effects on the natural environment.

For example, at the level of belief, religion and religious teachings shape the views adherents have of nature. Tanner and Mitchell (2002, p. 209) note that 'most religions view the natural environment as a harmonious unity, conceived by an external and transcendent mind'. From this perspective, Gardner (2002, p. 35), emphasises the traditional and important role of rituals, traditions, and religious institutions 'in governing sustainable use of the natural environment', as they act as 'a sophisticated social and spiritual technology' that helps mould the moral and ethical bearings of people, including 'people [living] in harmony with the natural world'. Building on the intricate connections between nature and religion, several conceptual approaches regarding these connections have been developed, including bioethics, sacred environmentalism, and deep-ecology (Chandran and Hughes, 1997; Nelson, 1998; Saraswati, 1998; Sullivan, 1998; Apffel-Marglin and Parajuli, 2000; Chapple and Tucker, 2000; Dwivedi, 2000).

At a more physical level, the heightened intersection between religious values and the environment can be witnessed in places that are deemed important for the practice of religion. These are in many cases centres of religious faith that are considered sacred or deeply rooted in the human need for understanding their place in the cosmos (Jackson, 1995; Shackley, 2001; Tanner and Mitchell, 2002). The notion of the 'sacred' in relation to place can be conceptualized in two ways (see Olsen, 2019a). The first conceptualization, often referred to as the Eliadean (Eliade, 1959) or 'ontological' view of sacred space, posits that hierophanies or divine irruptions take place in a particular location, which makes this location the *axis mundi* for religious groups. At many sacred sites, religious leaders incorporate various elements of nature into a system of religious values, symbols, and religious rituals (Marshall, 1994; Chandran and Hughes, 1997; Narayanan, 1997; Prorok, 1997; Shinde, 2008). Places such as Mt Meru, Mt Kailash, and Mt Manasarovar, which are believed to be the abode of Hindu God Shiva, belong to this category of sacred places. In the other conceptualization, referred to as the 'situationalist' perspective, sacred sites are not an 'ontological given' (della Dora, 2015) but rather are defined as sacred through political, sociocultural, economic, and psychological processes. Sacred sites, therefore, are not sacred until someone or some group deems the site as sacred through a deliberate process of sanctification (Gottlieb, 2004; Bremer, 2006). As Tanner and Mitchell (2002, p. 122) note, 'A sacred space is not normally extraordinary in its characteristics, it is made so by religious choice'.

Regardless of which view of how sacred space is created, the imbuing of mythological legends, miracle stories, and testimonies of the reality of divine forces within the physical elements in a sacred place contributes to the creation and maintenance of sacred geography. Often, these sacred places include a network of sacred sites linked to natural features, such as rivers, trees, and lakes marked by human-built shrines dedicated to certain deities and containing sanctified artefacts and objects. These sacred places, demarcated by religious groups by a sacred boundary to separate the sacred from the secular (Kong, 2001), become places where pilgrimage and other religious rituals are performed. This boundary in some cases also serves as a route of circumambulatory pilgrimage practices (Vidyarthi *et al.*, 1979; Haberman, 1994; Singh, 1997). Thus, attributes such as 'sacred', 'divine', 'religious', and 'spiritual' become central to the environment of sacred places. These same sacred places also provide opportunities to cross from the material world to the divine world (Eck, 1981) and crossing is symbolized in pilgrimage travel to such places. Experiencing the 'spirit of place' at these sacred sites attracts both believers and non-believers, and in the process they become epicentres of pilgrimage travel. Pilgrimage, then, becomes the religious framework through which the sanctity of the place is made accessible.

Yet, important as the relationships between religion and the environment are, there is a paucity of literature discussing these relationships within the context of everyday and mundane issues related to tourism development.

Sacred Spaces and Places: Where Religion, Tourism, and the Environment Intersect

Pilgrimage and religious tourism destinations are places where the intersections between religion, tourism, and the environment are significant, heightened, and more readily observed. While there are thousands of places that faith traditions and individuals consider sacred, not all of these places become popular pilgrimage sites (Stoddard and Morinis, 1997; Tanner and Mitchell, 2002). This is because, as noted above, a sacred site is sacred because of the visitation and ritual practices of religious adherents and visitors, wherein rituals are invested with religious beliefs, meanings, and imagery, and institutionalised in the activity of pilgrimage (Bharati, 1963; Turner and Turner, 1978; Morinis, 1984; Nolan and Nolan, 1989). These rituals and the visitation of these sites reinforce the sacred nature of the sites to pilgrims (Bremer, 2004). This positive feedback loop leads to 'visitors never doubt[ing] that they are experiencing a religious place, regardless of whether or not they share the religious proclivity of the place' (Bremer, 2000, p. 423).

The Physical Reality of Environmental Impacts

Sites of pilgrimage and religious tourism have abstract, metaphysical, and 'otherworldly' aspects to their existence. At the same time, they are physical places that are inhabited and act as a destination for religiously motivated visitors. These visitors, as Shackley (2001, p. 54) wrote, 'will have some impact, whether they wish or not'. Several scholars have discussed the many kinds of direct and indirect environmental impacts that occur in pilgrim-towns and within natural sacred landscapes due in part to religious tourism (Singh, 2002; Dasgupta *et al.*, 2006; Shinde, 2007a; Terzidou *et al.*, 2008; Verschuuren *et al.*, 2010; Alipour *et al.*, 2017). For example, based on observations in European cathedrals and churches, Shackley (2001) classified the direct impacts caused by visitors as *deliberate* (e.g. theft and vandalism), *thoughtless* (e.g. litter, pollution, noise), and *accidental* (e.g. abrasion of artwork or fabric in passing). These types of impacts are commonplace for religious buildings and natural landscapes that receive religious tourists (Nolan and Nolan, 1992; Swatos Jr. and Tomasi, 2002).

However, such impacts can be viewed very differently when considering religious tourism as an active religious practice, as is the case in many non-western societies. The scale of visitor impact is most evident in large gatherings around religious festivals, such as the *Kumbha Mela*,[1] a major pan-Indian pilgrimage event and the world's largest gathering of Hindu devotees that at times has 120 million participants. As a part of the *Kumbha Mela*, pilgrims participate in ritual bathing in the River Ganga. With millions of people participating in this ritual bathing, several environmental impacts occur. For example, a report prepared by the Central Pollution Control Board in India summarizes the environmental impacts of such mass bathing:

> Mass bathing is accompanied by mass defecation. Apart from that, the offerings of a plethora of materials – from ghee to flowers – are made to the river. This contributes high levels of organic matter to the river. As several infections are transmitted through water [...] there are good chances that the bathers are infected by viruses and pathogens that cause diseases like typhoid, cholera, bacterial dysentery and jaundice (cited in Ahmed *et al.*, 2000: para 20).

While this example of the *Kumbha Mela* may seem like a larger-than-life instance of the environmental impacts of religious ritual participation, the larger the group of participants in religious rituals, the worse the environmental impacts will be. Indeed, direct impacts are most visible at sacred destinations during festivals and events considered auspicious in religious faiths (Shinde, 2007a; Ruback *et al.*, 2008). These occasions, which closely follow religious calendars, may take place on a single day or occur over the space of several days, weeks, or even months (Singh, 1997). The continual influx of large numbers of pilgrims, let alone religious tourists, puts a severe strain on basic services such as water supply, sanitation, and waste management at these destinations (Kaur, 2019). The physical environment is further stressed with soil and water pollution and the clearing of land for the creation of temporary accommodation facilities and amenities for pilgrims (Nagabhushanam, 1997; Basheer, 2003).

In many cases, religious rituals, as significant expressions of individual or collective piety, involve the use of different kinds of material offerings, as noted above, which offerings have great potential for causing or adding to already existent environmental pollution (Ruback *et al.*, 2008). For example, many rituals in Hindu pilgrimage involve the sacrificial offerings of money and other material offerings, such as food and floral arrangements, and as pilgrims leave these religious offerings at sacred destinations they become material waste that further compounds the problem (see Shinde, Chapter 3, this volume). These additional waste pressures translate into unhygienic conditions and pose a major challenge for environmental sustainability (Sullivan, 1998; Sofield and Brent, 2001).

Changes in the seasonal nature of religious events have also exacerbated environmental impacts. It used to be that major religious festival occasions drew the largest influx of visitors and therefore the greatest intensity of environmental impact, while in between these major festivals smaller religious celebrations limited to local participation would occur with smaller environmental impacts (Picard and Robinson, 2006). However, not only have the number and frequency of these large religious celebrations and festivals increased (Shinde, 2007a, 2017; Shinde and Pinkney, 2013), in recent years, religious tourists have begun to visit these sacred destinations between the major religious festivals and events. As such, there are more people in these places for longer durations of time, meaning that the religiously-induced environmental problems that would normally be limited in scale and scope to these large religious events continue during the pilgrimage 'off-season'. This creates a situation where the resilience of religious sites are severely tested, as there is not enough time for the natural and built environment, let alone the host community, to recover from the impacts of major pilgrimage events. Therefore, there is a persistence of negative environmental impacts for longer durations, leading to a lack of 'seasonality' that is often inherent in leisure tourism.

While the direct environmental impacts, as the examples above note, are *directly* related to the presence of visitors, indirect environmental impacts are *induced* in many ways by the types of development patterns experience within these sites. The most significant environmental change in sacred destinations is urbanization that is driven by pilgrimage and tourism development, within which two processes are evident. The first process is driven by the consumptive needs of pilgrims, religious tourists, and leisure-oriented tourists. The focus on 'the consumption of pleasure' in these destinations creates a 'distinctive ecology, one evoking powerful images of pleasure – and this acts as a lure to [pilgrims and] tourists' (Mullins, 1991, p. 340). This 'distinctive ecology' leads to the rapid growth of tourism infrastructure, including hotel accommodation and transportation systems, to service increased visitor flows (Kaur, 1984; Singh, 2002). Also contributing to this rapid growth are the creation of new commercial establishments and real estate development. As Melwani (2001) notes in the pilgrimage centre of Haridwar-Rishikesh,

> Developers have constructed over 200 upscale apartment blocks, and many more are under construction ... Shivalik Ganga Estates in Haridwar uses the tag line: 'Peace Unlimited. Plots Limited.' Prices are upward bound – an apartment on the riverfront, which sold for US $25,500 4 years ago, now fetches $53,200. These luxury apartments may cost a bundle, but certainly the path to finding God was never easier!

As a part of this kind of new real estate development, there has been an increase in the purchasing of 'second homes' at sacred destinations (Shinde, 2008). The motivation to have a 'second-home' or a 'spiritual retreat' in 'god's abode' is compelling for devotees who are able and willing to pay a lot of money to own a piece of a sacred landscape (Shinde, 2017). This strong driver of sacred ownership creates a major incentive for the establishment of speculative real estate development.

The second process relates to the in-migration of people looking for work within the pilgrimage/tourism industry. Many of those who come to sacred destinations looking for work are poor migrants from neighbouring rural areas and small towns. Most of these migrants end up living in informal squatter settlements in urban peripheries that have no infrastructure services, which contributes to degradation of these peripheral areas of pilgrimage centres (Dube, 1994; Nagabhushanam, 1997). In addition, the lack of a pilgrimage/tourism off-season also fuels the demand for additional pilgrimage/tourism services and infrastructure, and attractions, leading to the creation of new urban landscapes that affect the resistance and resilience of the natural and human-built environment as well as the necessary physical infrastructure (Shinde, 2017).

These parallel processes lead to 'rapid population and labour force growth, a "flexible"' labour market, a [change in traditional] class structure, [and] a different household and residential organization' (Mullins, 1991, p. 340). As noted above, these processes also lead in many cases to the irreversible transformation of the sociospatial landscape of sacred destinations (Mullins, 1994; Gladstone, 1998), such as placing a higher than normal stress on the existing water supply system, sewage system, roads, and air quality (Varady, 1989; Srikrishna, 1994; Nagabhushanam, 1997; Trivedi and Agrawal, 2003). Thus, within growing sacred destinations, religious and leisure-oriented tourists seem to almost parasitically feed off the sacred core (Shinde, 2017).

Religious Perceptions of Pilgrimage and Religious Tourism Impacts

Many studies have found that visitors are often not concerned about their impacts on the natural or human-built environment in the sacred places they frequent despite being presented with evidence that they contribute to environmental degradation and unhygienic conditions (e.g. Alley, 1994; Sullivan, 1998; Haberman, 2000). Moreover, many of the physical environments within religious tourism destinations often present a stark contrast between the idealized sacral nature of the place and the reality of everyday living (Haberman, 2006). Turnbull (1981, p. 76) notes that in the case of Varanasi, while Hindu pilgrims perceive the city as the most beautiful and hallowed place in India, in reality the city 'may forever remain merely one of the dirtiest' for foreign visitors. Although more than three decades old, Turnbull's observation is still valid, as Varanasi continues to be one of India's most polluted yet sacred Hindu pilgrim-towns. In this vein, Tanner and Mitchell (2002, p. 125) point out that there is more to understanding the natural and human-built environments of sacred and religious places than through just scientific assessments and the quantification of environmental problems. They note that the power of these sacred places derives from their religious importance, and hence devotees are less inclined to focus on their degraded physical conditions or environment:

> Religious experiences which come from arriving in the sacred space are not related to the glories of nature or cultural aesthetics ... Whether at Lourdes or at the Buddhist pilgrimage centre at Katagarama in Sri Lanka, there is little visible at the site to inspire the pilgrim ... The reason must be sought in the mind and heart of individuals rather than in the objective value of places themselves.

People's perceptions of the environment are often shaped by cultural subjectivity. As Tuan (1974, p. 246) notes, 'culture can influence perception to a degree that people will see things that do not exist'. To better understand place-based meanings, McDowell (1994, p. 154) suggests that it is necessary to analyse both 'material practices and symbolic representations'. In this case, the issues and concerns around changes in the 'religious environment' can be raised regarding changes in religious and cultural traditions rather than problems related to the physical environment. Such articulation falls in line with McDowell's (1994, p. 146) observation that 'symbols, rituals, behaviour and everyday social practices result in a shared set or sets of meanings that are, to great or lesser degrees, place specific'.

For example, many pilgrimage sites are associated with natural features, including rivers, mountaintops, water bodies, and trees (Bhardwaj, 1997). Because these features are a part of a sacred geography defined by religious values and beliefs, in many cases they tend to be viewed from a religious perspective rather than in terms of their ecological functions (Eck, 1982; Gesler and Pierce, 2000, p. 222). Indeed, sacred places, even if they also function as a tourism destination, are not ordinary or secular environments, as they have undergone a process of sanctification and, as noted above, possess a 'spirit of place' that defines environment and identity. As such, while environmental change in these destinations may well be recognized and acknowledged, the followers of the faith for whom that sacred place is central and a manifestation of metaphysical views and beliefs may perceive this change differently. Both pilgrims and religious tourists therefore tend to overlook their role in the degradation of the environment (see Qurashi, Chapter 10, this volume), believing that their religious performances and positive economic contributions outweigh their impact (Terzidou *et al.*, 2008; Shinde, 2011). This is also the case at times with local residents, who may place the blame for environmental impacts on sources other than pilgrims. Shinde (2011, p.

456) highlights this perspective in an interview he had with a young guru in Vrindavan, who stated: 'Pilgrimage is a way to partake the spiritual energy of the place ... Visitors come with a religious belief and an emotional feeling ... How can they cause problems for the environment[?] ... What is a pilgrimage site without its visitors?'. Moreover, in pragmatic terms, visitors are critically important for maintaining the economy of the pilgrimage centre, and therefore residents may tolerate the negative impacts of tourism if it produces substantial economic benefits (Baedcharoen, 2000; Terzidou *et al.*, 2008). Thus, environmental problems cannot be solely placed on pilgrims' movement and actions, but is also based on perceptions about environmental degradation, especially when viewed through the lens of religion (Alley, 2002; Haberman, 2006).

Such perceptions, however, are not just theological or philosophical in nature, but are also generated through people's experiences with sacred places and how they view those who visit and engage with these same sites (Joseph, 1994; Lochtefeld, 2010). For example, Shoval (2000) argued that the staging of religious rituals for tourist consumption effectively devalued and commodified the sanctity of the city. Shackley (1999) also noted that the manipulation of religious rituals in Nepal to cater to tourist itineraries not only devalued the religious rituals, but also made it difficult for locals to participate in their own religious heritage. Hobbs (1992) noted that at Mount Sinai, Egypt, the development of tourist accommodation and activities had altered the sacred nature of the mountain, both in and around the monastery. Hobbs argued that the pressure to cater to tourists has resulted in many of the religious rituals at the monastery being replaced by staged performances, diminishing the role of the monks who previously interacted with and provided religious guidance to pilgrims. These impacts led Hobbs to lament that 'The physical, social and spatial requirements of caring for too many of the *wrong* kind of people—tourists—are weakening the fabric of monastic society at Mt. Sinai' (p. 99, emphasis added).

Indeed, at many sacred sites, tourists are seen as the 'wrong' type of people who, by their mere presence, undermine the religious significance and sanctity of the site (Joseph, 1994; Joseph and Kavoori, 2001; Shackley, 2004; Aggarwal *et al.*, 2008). In the Hindu pilgrim-town of Pushkar, Joseph (1994) notes that many traditional and orthodox religious communities consider both tourists and the religious intermediaries who have transformed *puja* (i.e. worship) rituals reserved for Hindus into commercial acts and touristic performances as actors that defile and pollute the sacred space. These communities also perceive the construction of hotels and tourist infrastructure as fundamentally altering the sacred landscape and the destinations' sense of place. This view of tourists and by extension the tourism industry has led to the view that tourism is solely responsible for the high pollution levels of the nearby Pushkar Lake.

In this same vein, while tourist behaviour reflects a certain disregard for religious protocol and a more hedonistic attitude towards sacred places, pilgrim behaviour is seen as being determined by religious protocol and therefore reflective of a sensitive relationship and sacred tie to the pilgrimage site (Cohen, 1992; Olsen and Timothy, 2006). Vukonić (1996), in describing some commonly-held perceptions regarding pilgrims and tourists, notes that 'The pilgrim steps gently on to holy soil; the tourist overruns holy places and photographs their remains. The pilgrim travels with humility and patience; the tourist travels arrogantly and in a hurry.'

Discourse, Materiality, and Management

The importance of understanding religious values as they shape perceptions about environmental problems in sacred places cannot be emphasized enough, particularly because they affect how sacred destinations are managed. The management of religious tourism sites has increasingly become an area of academic inquiry, highlighting the importance of quality visitor infrastructure and interpretational services as ways to minimize negative impacts on the natural and human-built environment (e.g. Olsen, 2006a; Raj and Morpeth, 2007; Zamani-Farahani and Henderson, 2010; Henderson, 2011; Shinde, 2012a; Bond *et al.*, 2015). However, even the implementation of best practices in crowd management can be overwhelmed in instances where the scale and frequency of visitation is such that no amount of management and control can stop negative impacts from occurring.

Some scholars argue from a public administration perspective that the real problem is the absence of a strong formal governance structure focusing specifically on pilgrimage and religious tourism and its environmental impacts (van Der Veer, 1988; Alley, 2002; FoV, 2006; Shinde, 2011; Raj and Griffin, 2015). As Tanner and Mitchell (2002, p. 204) note, environmental management is 'a political activity with little stress on religious considerations [,] even in states which are officially [claimed as religious] such as Israel, Saudi Arabia and Pakistan'. Also, religious views of pilgrimage and its (non)impact on the natural and human-built environment can lead to inconsistencies and disagreements over defining environmental issues, often leading to inaction by both government and religious authorities towards minimizing environmental impacts at pilgrimage and religious tourism destinations. Discussions regarding environmental change at pilgrimage destinations are often debated by contrasting the material reality of the environment with the abstract notion of sacred space. For instance, several scholars have noted the tendency of Hindu pilgrims and religious functionaries to compare the contemporary state of the environment of pilgrimage sites with a glorious past rooted in mythology or in the concept of transcendence (Haberman, 1994; Lochtefeld, 2010). This comparison relies on a fixed notion of the other worldly divine quality of sacred space that reinforces the sanctity of a pilgrimage site while outright ignoring the negative impacts of pilgrims and pilgrimage practices on the physical environment. This discourse of transcendence therefore produces conflicting views regarding who is responsible for managing the natural and human-built environment of a sacred place (Freitag, 1989; Alley, 2002). Religious actors often invoke tradition to justify their use of environmental resources for religious practices while claiming no responsibility for and management of its impacts. Such dissonance produces an environment of neglect and apathy, shifting the responsibilities of environmental management to government entities that may not have the financial or technical resources to do so (Shinde, 2011).

Following the ethical–moral perspectives and inherent teachings of many religious traditions regarding respect for nature, a few studies have found that religious institutions that place a strong ethic and concern on environmental stewardship are more effective at addressing environmental degradation at their sacred sites (Grim and Tucker, 1994; Sullivan, 1998; Apffel-Marglin and Parajuli, 2000). One illustrative example is the Vrindavan Conservation Project in India. Vrindavan is a pilgrim-town dedicated to the Hindu god Krishna and is believed to be Krishna's recreational playground. As a part of this project, religious gurus have taken the lead to restore the forests in the name of Krishna and encourage other stakeholders to be more environmentally responsible (for more details see Prime, 2008; Nash, 2012; Chapter 4, this volume).

The Ecosystem of Religious Tourism: A Conceptual Model

The creation of a general theory regarding the intersections between religion, tourism, and the environment needs to move beyond the documentation of the environmental impacts of religious tourism which, although necessary, remains incomplete. In Fig. 1.2, the authors propose a simplistic conceptual model that can help understand the ecosystem of religious tourism that occurs in sacred places.

Sacred places are specific environments that connect elements of nature with sacred values using belief systems, mythology, cosmology, history, and culture involving religious faiths. The material manifestation of places imbued with such values generates a unique spirit of place

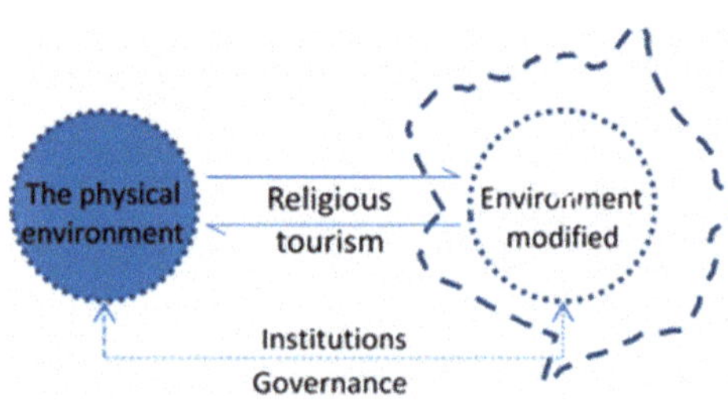

Fig. 1.2. A conceptual model of the relationships between religion, tourism, and the environment.

and a natural, sociocultural, and physical environment. Both the physical and metaphysical nature of these places are articulated through religious frameworks and their meanings experienced through rituals and performances. As such, travelling to engage with the spirit of these places becomes the focus of religious tourism. Travel in and of itself becomes the cause of change in the physical and spiritual nature of these places. How the environment is impacted and changes depends largely on the nature of religious mobilities and the ways in which religious and governmental institutions mediate experiences with the sacred. This process is dialectical and leads to a continuous evolution of the people who visit, the people that live there, and the institutions that manage the place.

This theoretical model can be illustrated using a setting of a generic pilgrim-town. In this town, there are several institutions that interact with each other to produce and reproduce the ecosystem within which religious tourism operates. These institutions can be broadly classified into five groups, which are not exclusive of each other and can have significant overlap between them. The first, and probably most significant, includes leaders of religious establishments – the gurus, and the ritual priests that form the core of the pilgrimage industry and are central to the catering for the religious and spiritual wants and needs of pilgrims. These 'religious entrepreneurs' seek to establish and maintain relationships of patronage with individuals, families, or community groups that invest in religious buildings and charities and engage in ritual performances. Second, the pilgrimage/religious tourism industry is also comprised of several people engaged in supplementary trades that provide goods and services to visitors. Their activities depend on broader socioeconomic processes and factors, such as the economic status of patrons and pilgrims, patterns of visitation, modes of travel, availability and types of transportation, and the frequency of travel, among others.

Third, residents form an important institution within this pilgrim-town. Although religious entrepreneurs and tradespersons may form part of this category, residents also include devotees and other social groups that may serve religious establishments and businesses or work in other sectors of an urban economy. The fourth group of institutions are those involved in the governance and maintenance of the pilgrimage-town. These institutions may include state agencies operating across different scalar levels, such as municipalities and state government departments, as well as other corporate religious bodies and civic associations that are often classified as non-governmental organizations (NGOs). One notable example of this is the Sathya Sai Baba Trust in Puttaparti in Andhra Pradesh, India, that was established in 1972 by a highly revered guru named Sri Sathya Sai Baba. This trust has contributed significantly to the building of infrastructure and environmental services in the town of Puttaparti. Consisting of more than 10,000 employees, the trust runs an independent university and implements charitable projects providing social welfare in the region. One of its much discussed environmental projects is the water supply project that it has developed in partnership with the state government and a private sector company to provide drinking water in 750 villages in the Andhra Pradesh state (Parthasarthy, 2002). The fifth institutional group are the pilgrims that visit the pilgrim-town. Depending on their economic status, some pilgrims may exercise substantial influence on the pilgrimage centre through their patronage of religious establishments and funding of charities, temples, and public services for fellow travellers. Religious tourists can also be included in this group because of their importance to the continuing economic growth of these destinations in both the religious and secular sectors.

The creation and growth of any pilgrimage site depends on the extent to which these institutions interact in a positive manner to reinforce the religious importance of the place. The ritual practices performed by religious intermediaries, the services and artefacts provided by supporting trades, and the symbolism of various religious events and festivals sustain the religious identity and sacrality of sacred places. But pilgrimage, as both ritual practice and economy, is not independent from the changes in the broader socioeconomic and political realm within which it occurs and is therefore likely to change and consequently transform the religious environment (Shackley, 2001). These changes, as noted above, may be visible in the reorganisation of ritual performances, changes in seasonal patterns of pilgrimage flows, and the characteristics of visitors and religious investments that both shape, and are shaped by, the physical and built environment of the pilgrimage centre. Transformations in the socioeconomic and physical environment of pilgrimage centres are also shaped by changing representations of religiosity, piety, and sacredness of the pilgrimage landscape. These representations are articulated by religious institutions through discourses that interpret the changing environment of pilgrimage in different ways. Some may invoke mythological stories or past traditions to criticize or challenge the changes occurring in the built environment, while others might claim that the degradation of the physical environment reflects moral degeneracy and defilement of the sacred site. Other still might embrace religious symbols to argue for the modernization of religious traditions and pilgrimage sites. Together, these competing discourses of change produce new institutional spaces and configurations for regulating, managing, and sustaining the material landscape and cultural symbolism of the place.

Chapter Contributions

This quick review of the existing literature around the key concepts of tourism, religion, and environment shows that while each of these areas has its own exploratory domains, the relationships between these concepts have received varying degrees of attention. Of these connections, tourism—environment relations have received the greatest amount of attention and yielded basic conceptual frameworks to understand these relationships. However, the role of religious values that influence both positive and negative impacts between tourism and the environment are largely missing. The connections between tourism and religion only sparingly discuss environmental aspects, and regarding the relationships between religion and environment, the appeal of nature and ethical calls for environmental stewardship are quite evident, but investigations into the actual interactions of the various aspects that underpin these relationships within the material world are still lacking.

The chapters included in this volume elaborate on some of the core aspects of this theoretical model presented in this chapter that views 'the environment' of pilgrimage as a dynamic process shaped by the activities, forms of control, perceptions, and representations of the actors involved in the production of sacred sites. Some chapters focus on the environment as a resource or generator for religious tourism, while other chapters discuss the environment as a recipient of impacts of religious tourism.

In addition to this introductory chapter, there are eleven contributions in this book. In Chapter 2, Daniel Olsen examines the relationships between pilgrimage, religious tourism, biodiversity, and natural sacred sites, and how religiously motivated travel to these sites affects the natural environment. After highlighting the views of nature by several major faith traditions, Olsen turns to examine how natural sacred sites are defined and categorises these sites into sacred mountains and volcanoes, caves and grottoes, water, trees and groves, and plants and animals. The author then notes some of the impacts of pilgrimage and religious tourism on natural sacred sites, and finishes by discussing attempts at different scales to maintain the sanctity and sustainability of these sites through governance frameworks that focus on the preservation of biodiversity.

In Chapter 3, Kiran Shinde engages in a comparison of six religious tourism destinations in India through the analytical categories of environmental processes, institutional responsibility, place attachment, and governance systems. In doing so, Shinde expands on the conceptual model in this chapter (Fig. 1.2) to provide a comprehensive explanation of the complexities of managing the natural and human-built environment at religious tourism destinations.

In Chapter 4, Joshua Nash focuses on how deeper and more philosophical experiences with sacred places, such as Vrindavan, India, can lead to the creation of human sanctuaries. Although present-day Vrindavan, which in popular mythological image is considered a celestial forest and divine playground of Krishna, is a city full of environmental problems, it still is the focus of spiritual teachings that build on the union of religion and nature conservation. In an almost biographical style, Nash expounds on the Vaishnava teachings and their relevance for conserving the natural environment as he learned them from his study with local Vrindavan ecologist Shri Sevak Sharan, who calls for practical action and contemplative practice (*sadhana, chintan*) using what he calls the Vrindavan Ecological Concept (VEC). Nash argues that this concept is powerful enough to inspire changes in thought processes and behavioural action that can create human sanctuaries everywhere in the world.

In Chapter 5, Sanjay Nepal, Yang Mu, and Po-Hsin Lai examine how local residents make sense of the sacred landscape of the Sagarmatha (Mt Everest) National Park (SNP) in Nepal's Khumbu Region. From their fieldwork they report that Sherpas' emotional and spiritual bonds with the landscape are reinforced through routinized contact with the sacred cultural features and consistent compliance with the rules of conduct. The authors argue that Sherpa residents' objective recognition of the sacred landscape is influenced by geographical proximity to sacred sites, age of residents, and their religious backgrounds. Subjective interpretations of sacred landscape are also shaped by a strong awareness of behavioural restrictions, family influence and personal experiences. These strong bonds, however, are under threat as the religious influence of the landscape is diminishing with the shift of people's economic dependency from the land to a consumer culture facilitated by tourism.

In Chapter 6, Anouk Lafortune-Bernard, Rajendra Suwal, Kai Weise, and Robin Coningham discuss the challenges for site managers at the Limbini World Heritage Site in Nepal, the birthplace of the Buddha. Since 1978, the Lumbini Master Plan has focused on ecological conservation and preserving the sanctity of the site in the face of increasing urbanization and tourism development in the area. The authors reflect on some of the difficulties in implementing the original plan, showing that the lack of integration of the Lumbini Master Plan with other local and regional plans has made it difficult to achieve the broader sustainable goals for the region.

In Chapter 7, Nour Farra-Haddad looks at interfaith pilgrimages in Lebanon where Christian and Muslims pilgrims worship at each other's sacred sites. More specifically, the author looks at how the natural elements of these sacred sites, including trees, water, and stones, are utilized by pilgrims. In doing so, Farra-Haddad argues that the use of these natural elements in shared religious rituals acts as a catalyst for interfaith cooperation and experiences between different faith traditions in Lebanon.

The importance of understanding the sociopolitical contexts in which a sacred site is constructed is the focus of Josep-Maria Garcia-Fuentes in Chapter 8. Garcia-Fuentes discusses the building of the Sagrada Familia, a well-known Catholic expiatory temple in Barcelona. The author argues that its design and construction was symbolic of competing religious and political forces to use the mountain and shrine of Montserrat as a national symbol. This chapter situates the Sagrada Familia by Gaudí and its entangled relations with Montserrat within the reconceptualization of nature by the Catholic Church in a contemporary European context.

In Chapter 9, Daniel Olsen examines how the rigours of religious tourism and pilgrimage affect human health through a review of the literature related to the development and spread of infectious diseases by participants at pilgrimage/religious tourism mass gatherings. While some of this occurs because of the lack of hygienic facilities, the close proximity of religious travellers in sometimes small, confined spaces also exacerbates the rates of contraction and the spread of these diseases. The author also looks at the different types of infectious and non-infectious disease and other health risks that occur at religious mass gatherings with a particular emphasis on the *Hajj*. The author concludes with a discussion of how government, health, and religious officials are trying to mitigate the spread of diseases that originate from, and other health-related issues that occur at, these religious mass gatherings, highlighting the need for further research in this area.

Waste management, the most visible environmental problem for Hajj managers, is the focus on Jahanzeeb Qurashi's chapter (Chapter 10). Even though the Quran is explicit in highlighting the importance of environmental stewardship and conservation, the areas in Saudi Arabia that host the *Hajj* – one of the most important tenets of Islamic religious practice – experience extensive environmental damage during the pilgrimage. This is in part due to the lack of controlled waste disposal, recovery, and recycling, as well as deficiencies in the country's national waste management strategy. This chapter critically focuses on a number of key areas related to the *Hajj* and environmental damage, including how pilgrims and the hospitality and tourism industry affect the natural environment and how the government of Saudia Arabia is attempting to improve the environmental sustainability of the *Hajj*.

In Chapter 11, Michael Di Giovine and Elisa Ascione focus on the rebuilding of religious tourism sites after major natural disasters, as shown in the case of St Benedict in Norcia, Italy. The authors explore how different stakeholders have framed the role of tourism in the area and how touristic practices re-signify the value of places and their associated moral ideals after environmental disasters. While the clergy in the area have used recent earthquakes as a metaphor for redemption and renewal, tourists and other stakeholders have also affirmed their presence and experiences in the region as transformative for themselves and for the environment. The authors also emphasize how the necessary tearing down of important churches affects the region's sense of place and social identity, and how efforts to rebuild these churches as well as the surrounding community have helped instil a sense of sacredness beyond the old core to the destination as a whole. In addition, the authors illustrate the dialectical effects of a double environmental transformation on sacred sites: how a natural disaster – unplanned and uncontrollable by social actors – affects the tangible and intangible fabric of a pilgrimage site and how those subsequent social interventions likewise influence religious tourism and its stakeholders as well as the wider environment.

In the final chapter, Daniel Olsen and Kiran Shinde summarize the major arguments as discussed in the preceding chapters and expand upon their proposed conceptual framework for environmental analysis in a pilgrimage place from Chapter 1. The authors also focus on the limitations of the book and suggest future avenues of research in the area of religion, tourism, and the environment.

Note

[1] The *Kumbha Mela* celebrates the victory of the gods over the demons in winning control over the pitcher or kumbha containing the nectar of immortality. According to myth, four drops of the nectar fell to earth during the battle, each landing in one of four religious centres: Allahabad, Haridwar, Ujjain, and Nasik. The *Kumbha Mela* pilgrimage, which takes place every 3 years in January-February, rotates between these four religious centres. Participating in the *Kumbha Mela* is regarded as a meritorious act for Hindus, and *sadhus* or wandering ascetics, religious gurus, and pilgrims from all over India and the world come to participate in the ritual bathing in the river on the day marking the mythical event. In 2019, more than 120 million visitors attended the *Kumbha Mela* at Allahabad.

References

Aggarwal, A.K., Guglani, M. and Goel, R.K. (2008) Spiritual & yoga tourism: A case study on experience of foreign tourists visiting Rishikesh, India. *Proceedings of the Conference on Tourism in India – Challenges Ahead*, IIMK, 15-17 May.

Ahmed, K., Kashyap, S. and Sinha, S.K. (2000) Pollution of Hinduism. Down to Earth. Available at: https://www.downtoearth.org.in/coverage/pollution-ofhinduisim-17622 (accessed 21 January, 2020).

Alipour, H., Olya, H.G.T. and Forouzan, I. (2017) Environmental impacts of mass religious tourism: from residents' perspectives. *Tourism Analysis* 22(2), 167–183. DOI: 10.3727/108354217X14888192562285.

Alley, K.D. (1994) Ganga and Gandagi: interpretations of pollution and waste in Banaras. *Ethology: formerly Zeitschrift für Tierpsychologie* 33, 127–145.

Alley, K.D. (2002) *On the Banks of the Ganga: When Waste Water Meets a Sacred River*. University of Michigan Press, Ann Arbor, Michigan.

Ambrósio, V. (2003) Religious tourism: territorial impacts in the EU Marian sanctuary towns. In: Fernandes, C., Mcgettigan, F. and Edwards, J. (eds) *The 1st Expert Meeting of ATLAS- Religious Tourism And Pilgrimage Research Group*. Tourism Board of Leiria, Fatima, pp. 43–56.

Andriotis, K. (2009) Sacred site experience: a phenomenological study. *Annals of Tourism Research* 36(1), 64–84.

Andriotis, K. (2011) Genres of heritage authenticity: Denotations from a pilgrimage landscape. *Annals of Tourism Research* 38(4), 1613–1633.

Apffel-Marglin, F. and Parajuli, P. (2000) Sacred grove and ecology: ritual and science. In: Chapple, C.K. and Tucker, M.E. (eds) *Hinduism and Ecology: The intersection of Earth, Sky and Water*. Harvard University Press for the Center for the Study of World Religions, Harvard Divinity School, Cambridge, Massachusetts, pp. 291–312.

Badone, E. and Roseman, S. (2004) *Intersecting Journeys: The Anthropology of Pilgrimage and Tourism*. University of Illinois Press, Urbana and Chicago.

Baedcharoen, I. (2000) Impacts of Religious Tourism in Thailand. Masters Thesis. University of Otago, Otago, New Zealand.

Ballantyne, R. and Packer, J. (2013) *International Handbook on Ecotourism*. Edward Elgar Publishing, Northampton, Massachusetts.

Basheer, M.P. (2003) Faith in spate: Sabarimala bursting at the seams with devotees but environmental concerns ignored. Down to Earth. Available at: https://www.downtoearth.org.in/coverage/faith-in-spate-12356 (accessed 21 January 2020).

Bharati, A. (1963) Pilgrimage in the Indian tradition. *History of Religions* 3(1), 135–167. DOI: 10.1086/462476.

Bhardwaj, S.M. (1997) Geography and pilgrimage: a review. In: Stoddard, R.H. and Morinis, A. (eds) *Sacred Places, Sacred Spaces: The Geography of Pilgrimages*. Louisiana State University Press, Baton Rouge, pp. 1–24.

Bond, N., Packer, J. and Ballantyne, R. (2015) Exploring visitor experiences, activities and benefits at three religious tourism sites. *International Journal of Tourism Research* 17(5), 471–481. DOI: 10.1002/jtr.2014.

Bouyahya, D. (2016) Religious tourism and Sufism in Morocco. *European Journal of Multidisciplinary Studies* 2(1), 40–50. DOI: 10.26417/ejms.v2i1.p40-50.

Bremer, T.S. (2000) Tourists and religion at Temple Square and Mission San Juan Capistrano. *The Journal of American Folklore* 113(450), 422–435. DOI: 10.2307/542040.

Bremer, T.S. (2004) *Blessed with Tourists: The Borderlands of Religion and Tourism in San Antonio*. University of North Carolina Press, Chapel Hill, North Carolina.

Bremer, T.S. (2006) Sacred spaces and tourist places. In: Timothy, D.J. and Olsen, D.H. (eds) *Tourism, Religion and Spiritual Journeys*. Routledge, London and New York, pp. 25–35.

Buckley, R. (2011) Tourism and environment. *Annual Review of Environment and Resources* 36(1), 397–416. DOI: 10.1146/annurev-environ-041210-132637.

Butler, R.W. (1980) The concept of a tourist area cycle of evolution: implications for management of resources. *The Canadian Geographer/Le Géographe canadien* 24(1), 5–12. DOI: 10.1111/j.1541-0064.1980.tb00970.x.

Butler, R. (2006) *The Tourism Area Life Cycle Vol. 2: Conceptual and Theoretical Issues*. Channel View Publications, Clevedon, UK.

Butler, R. and Suntikul, W. (2018) *Tourism and Religion: Issues and Implications*. Channel View Publications, Bristol, UK.

Chandran, M.D.S. and Hughes, J.D. (1997) The sacred groves of South India: ecology, traditional communities and religious change. *Social Compass* 44(3), 413–427. DOI: 10.1177/003776897044003008.

Chapple, C.K. and Tucker, M.E. (eds) (2000) *Hinduism and Ecology: The Intersection of Earth, Sky and Water*. Harvard University Press for the Center for the Study of World Religions, Cambridge, Massachsetts.

Chen, L.I.-L., Scott, N. and Benckendorff, P. (2017) Mindful tourist experiences: a Buddhist perspective. *Annals of Tourism Research* 64, 1–12. DOI: 10.1016/j.annals.2017.01.013.

Cohen, E. (1978) The impact of tourism on the physical environment. *Annals of Tourism Research* 5(2), 215–237. DOI: 10.1016/0160-7383(78)90221-9.

Cohen, E. (1992) Pilgrimage and tourism: convergence and divergence. In: Morinis, A. (ed.) *Sacred Journeys: The Anthropology of Pilgrimage*. Greenwood Press, Westport, Connecticut, pp. 47–61.

Cohen Ioannides, M.W. and Ioannides, D. (2006) Global Jewish tourism: Pilgrimages and remembrance. In: Timothy, D.J. and Olsen, D.H. (eds) *Tourism, Religion and Spiritual Journeys*. Routledge, London and New York, pp. 156–171.

Collins-Kreiner, N. (2010a) Current Jewish pilgrimage tourism: modes and models of development. *Tourism* 58(3), 259–270.

Collins-Kreiner, N. (2010b) The geography of pilgrimage and tourism: transformations and implications for applied geography. *Applied Geography* 30(1), 153–164. DOI: 10.1016/j.apgeog.2009.02.001.

Collins-Kreiner, N. and Kliot, N. (2000) Pilgrimage tourism in the Holy land: the behavioural characteristics of Christian pilgrims. *GeoJournal* 50(1), 55–67. DOI: 10.1023/A:1007154929681.

Collins-Kreiner, N. and Luz, N. (2018) Judaism and tourism over the ages: the impacts of technology, geopolitics and the changing political landscape. In: Butler, R. and Suntikul, W. (eds) *Tourism and Religion: Issues and Implications*. Channel View Publications, Bristol, pp. 51–67.

Damari, C. and Mansfeld, Y. (2016) Reflections on pilgrims' identity, role and interplay with the pilgrimage environment. *Current Issues in Tourism* 19(3), 199–222. DOI: 10.1080/13683500.2014.957660.

Dasgupta, S., Mondal, K. and Basu, K. (2006) Dissemination of cultural heritage and impact of pilgrim tourism at Gangasagar Island. *The Anthropologist* 8(1), 11–15. DOI: 10.1080/09720073.2006.11890928.

della Dora, V. (2015) Sacred space unbound, *Society and Space*. Available at: http://societyandspace.org/2015/06/12/virtual-issue-13-sacred-space-unbound/ (accessed 21 January 2020).

Din, K.H. (1989) Islam and tourism: patterns, issues, and options. *Annals of Tourism Research* 16(4), 542–563.

Doron, A. (2005) Encountering the 'other': Pilgrims, tourists and boatmen in the city of Varanasi. *The Australian Journal of Anthropology* 16(2), 157–178. DOI: 10.1111/j.1835-9310.2005.tb00032.x.

Dowling, R. and Newsome, D. (2018) *Handbook of Geotourism*. Edward Elgar Publishing, Northampton, Massachusetts.

Dube, K.K. (1994) Urban sprawl of Varanasi: problems and planning. In: Singh, H.H., Dube, K.K., Singh, S.B., Kumra, V.K. and Singh, J. (eds) *Patterns of Urban Change in India*. Indrasini Devi Publications, Varanasi, pp. pp. 215–.222.

Dwivedi, O.P. (2000) Dharmic ecology. In: Chapple, C.K. and Tucker, M.E. (eds) *Hinduism and Ecology: The Intersection of Earth, Sky and Water*. Harvard University Press for the Center for the Study of World Religions, Cambridge, Massachusetts, pp. 3–20.

Eck, D. (1981) India's tirthas: 'Crossing' in Sacred Geography. *History of Religions* 20(2), 323–344.

Eck, D. (1982) *Banaras, City of Light*. Alfred A. Knopf, New York.

El Hanandeh, A. (2013) Quantifying the carbon footprint of religious tourism: the case of hajj. *Journal of Cleaner Production* 52, 53–60. DOI: 10.1016/j.jclepro.2013.03.009.

Eliade, M. (1959) *The Sacred and the Profane: The Nature of Religion*. Harcourt Brace Jovanovich, New York.

Feldman, J. (2007) Constructing a shared Bible land: Jewish Israeli guiding performances for Protestant pilgrims. *American Ethnologist* 34(2), 351–374. DOI: 10.1525/ae.2007.34.2.351.

Fleischer, A. (2000) The tourist behind the pilgrim in the Holy land. *International Journal of Hospitality Management* 19(3), 311–326. DOI: 10.1016/S0278-4319(00)00026-8.

FoV (2006) *Vrindavan Garbage Management Initiative*. Friends of Vrindavan, Vrindavan.

Freitag, S.B. (ed.) (1989) *Culture and Power in Banaras: Community, Performance, and Environment, 1800-1980*. University of California Press, Berkeley, California.

Gardner, G. (2002) *Invoking the Spirit: Religion and Spirituality in the Quest for a Sustainable World*. World watch Institute, Washington, DC.

Gatrell, J.D. and Collins-Kreiner, N. (2006) Negotiated space: Tourists, pilgrims, and the Bahá'í terraced gardens in Haifa. *Geoforum* 37(5), 765–778. DOI: 10.1016/j.geoforum.2006.01.002.

Geary, D. and Mason, D. (2016) Walking in the Valley of the Buddha: Buddhist revival and tourism development in Bihar. *Journal of Global Buddhism* 17, 57–63.

Gesler, W.M. and Pierce, M. (2000) Hindu Varanasi. *Geographical Review* 90(2), 222–237. DOI: 10.1111/j.1931-0846.2000.tb00332.x.

Ghosal, S. and Maity, T. (2010) Development and sustenance of Shirdi as a centre for religious tourism in India. In: Singh, R.P.B. (ed.) *Holy Places and Pilgrimages: Essays on India*. Shubhi Publications, New Delhi, pp. 159–173.

Gladstone, D.L. (1998) Tourism urbanization in the United States. *Urban Affairs Review* 34(1), 3–27. DOI: 10.1177/107808749803400101.

Goldberg, K. (2013) Pilgrimage re-oriented: Buddhist discipline, virtue and engagement in Bodhgayā. *The Eastern Buddhist* 44(2), 95–120.

Gottlieb, R.S. (ed.) (2004) *This Sacred Earth: Religion, Nature, Environment*. Routledge, New York.

Grim, J. and Tucker, M. (eds) (1994) *Worldviews and Ecology: Religion, Philosophy, and the Environment*. Orbis Books, Maryknoll, New York.

Gupta, V. (1999) Sustainable tourism: learning from Indian religious traditions. *International Journal of Contemporary Hospitality Management* 11(2/3), 91–95. DOI: 10.1108/09596119910250751.

Gursoy, D., Kim, K. and Uysal, M. (2004) Perceived impacts of festivals and special events by organizers: an extension and validation. *Tourism Management* 25(2), 171–181. DOI: 10.1016/S0261-5177(03)00092-X.

Haberman, D.L. (1994) *Journey Through the Twelve Forests: An Encounter with Krishna*. Oxford University Press, New York.

Haberman, D.L. (2000) River of love in an age of pollution. In: Chapple, C.K. and Tucker, M.E. (eds) *Hinduism and Ecology: The Intersection of Earth, Sky and Water*. Harvard University Press for the Center for the Study of World Religions, Cambridge, Massachusetts, pp. 339–352.

Haberman, D.L. (2006) *River of Love in an Age of Pollution: The Yamuna River of Northern India*. University of California Press, Berkeley, California.

Hall, C.M. (2006) Buddhism, tourism and the middle way. In: Timothy, D.J. and Olsen, D.H. (eds) *Tourism, Religion and Spiritual Journeys*. Routeldge, London and New York, pp. 172–185.

Heintzman, P. (2009) Nature-based recreation and spirituality: a complex relationship. *Leisure Sciences* 32(1), 72–89. DOI: 10.1080/01490400903430897.

Heintzman, P. (2014) Nature-based recreation, spirituality and persons with disabilities. *Journal of Disability & Religion* 18(1), 97–116. DOI: 10.1080/15228967.2014.868983.

Henderson, J.C. (2009) Islamic tourism reviewed. *Tourism Recreation Research* 34(2), 207–211. DOI: 10.1080/02508281.2009.11081594.

Henderson, J.C. (2011) Religious tourism and its management: the hajj in Saudi Arabia. *International Journal of Tourism Research* 13(6), 541–552. DOI: 10.1002/jtr.825.

Hobbs, J.J. (1992) Sacred space and touristic development at Jebel Musa (MT Sinai), Egypt. *Journal of Cultural Geography* 12(2), 99–113. DOI: 10.1080/08873639209478413.

Holden, A. (2003) In need of new environmental ethics for tourism? *Annals of Tourism Research* 30(1), 94–108. DOI: 10.1016/S0160-7383(02)00030-0.

Holden, A. (2007) *Environment and Tourism*. Routledge, London and New York.

Holden, A. and Fennell, D.A. (2012) *The Routledge Handbook of Tourism and the Environment*. Routledge, London and New York.

Huntsinger, L. and Fernández-Giménez, M. (2000) Spiritual pilgrims at Mount Shasta, California. *Geographical Review* 90(4), 536–558. DOI: 10.1111/j.1931-0846.2000.tb00353.x.

Iso-Ahola, S.E. (1980) *The Social Psychology of Leisure and Recreation*. William C. Brown Co. Publishers, Dubuque, Iowa.

Jackowski, A. and Smith, V.L. (1992) Polish pilgrim-tourists. *Annals of Tourism Research* 19(1), 92–106. DOI: 10.1016/0160-7383(92)90109-3.

Jackson, R.H. (1995) Pilgrimage in an American religion: Mormons and secular pilgrimage. In: Dubey, D.P. (ed.) *Pilgrimage Studies: Sacred Places, Sacred Traditions*. Society of Pilgrimage Studies, Allahabad, India, pp. 191–208.

Joseph, C.A. (1994) Temples, Tourists and the Politics of Exclusion: the Articulation of Sacred Space at the Hindu Pilgrimage Centre of Pushkar, India. Ph.D. Thesis. University of Rochester.

Joseph, C.A. and Kavoori, A.P. (2001) Mediated resistance: tourism and the host community. *Annals of Tourism Research* 28, 998–1009.

Jutla, R.S. (2000) Visual image of the city: tourists' versus residents' perception of Simla, a Hill station in northern India. *Tourism Geographies* 2(4), 404–420. DOI: 10.1080/146166800750035512.

Jutla, R.S. (2002) Understanding Sikh pilgrimage. *Tourism Recreation Research* 27(2), 65–72. DOI: 10.1080/02508281.2002.11081222.

Jutla, R.S. (2006) Pilgrimage in Sikh tradition. In: Timothy, D.J. and Olsen, D.H. (eds) *Tourism, Religion and Spiritual Journeys*. Routledge, London and New York, pp. 206–219.

Kamarudin, L.M. and Nizam, H. (2013) Islamic Tourism: The Impacts to Malaysia's Tourism Industry. Proceedings of international conference on tourism development. Available at: http://eprints.usm.my/35030/1/HBP40.pdf (accessed 21 January, 2020).

Kaur, J. (1984) Badrinath: Himalaya's super pilgrimage resort. *Tourism Recreation Research* 9(2), 21–26. DOI: 10.1080/02508281.1984.11015140.

Kaur, B. (2019) Faith to filth: thanks to Kumbh, Prayagraj sinks in solid waste.. Down to Earth. Available at: https://www.downtoearth.org.in/news/pollution/faith-to-filth-thanks-to-kumbh-prayagraj-sinks-in-solid-waste-64579 (accessed 21 January 2020).

Kong, L. (2001) Mapping 'new' geographies of religion: politics and poetics in modernity. *Progress in Human Geography* 25(2), 211–233. DOI: 10.1191/030913201678580485.

Kosansky, O. (2002) Tourism, charity, and profit: the movement of money in Moroccan Jewish pilgrimage. *Cultural Anthropology* 17(3), 359–400. DOI: 10.1525/can.2002.17.3.359.

Kreiner, N.C., Shmueli, D.F., Ben Gal, M., Collins-Kreiner, N. and Gal, M.B. (2015) Understanding conflicts at religious-tourism sites: The Baha'i World Center, Israel. *Tourism Management Perspectives* 16, 228–236. DOI: 10.1016/j.tmp.2015.04.001.

Leiper, N. (1990) Tourist attraction systems. *Annals of Tourism Research* 17(3), 367–384. DOI: 10.1016/0160-7383(90)90004-B.

Liu, J.C., Sheldon, P.J. and Var, T. (1987) Resident perception of the environmental impacts of tourism. *Annals of Tourism Research* 14(1), 17–37. DOI: 10.1016/0160-7383(87)90045-4.

Lochtefeld, J. (2010) *God's Gateway: Identity and Meaning in a Hindu Pilgrimage Place*. Oxford University Press, New York.

Marshall, J. (1994) The mosque on erb street: a study in sacred and profane space. *Environments* 22(2), 55–65.

McDowell, L. (1994) The transformation of cultural geography. In: Gregory, D., Martin, R. and Smith, G. (eds) *Human Geography: Society Space and Social Sciences*. Macmillan, Basingstoke, pp. 146–164.

McIntosh, A.J. and Prentice, R.C. (1999) Affirming authenticity: consuming cultural heritage. *Annals of Tourism Research* 26(3), 589–612.

Melwani, L. (2001) The joys of pilgrimage, Hinduism Today [Online], 12. Available at: https://www.hinduismtoday.com/modules/smartsection/item.php?itemid=4118 (accessed 21 January 2020).

Mesch, G.S. and Manor, O. (1998) Social ties, environmental perception, and local attachment. *Environment and Behavior* 30(4), 504–519. DOI: 10.1177/001391659803000405.

Morinis, A. (1984) *Pilgrimage in Hindu Tradition: A Case Study of West Bengal*. Oxford University Press, Delhi.

Mostafanezhad, M., Norum, R., Shelton, E.J. and Thompson-Carr, A. (2016) *Political Ecology of Tourism: Community, Power and the Environment*. Routledge, London and New York.

Moufakkir, O. and Selmi, N. (2018) Examining the spirituality of spiritual tourists: a Sahara desert experience. *Annals of Tourism Research* 70, 108–119. DOI: 10.1016/j.annals.2017.09.003.

Mullins, P. (1991) Tourism urbanization. *International Journal of Urban and Regional Research* 15(3), 326–342. DOI: 10.1111/j.1468-2427.1991.tb00642.x.

Mullins, P. (1994) Class Relations and Tourism Urbanization: The Regeneration of the *Petite Bourgeoisie* and the Emergence of a New Urban Form. *International Journal of Urban and Regional Research* 18(4), 591–608. DOI: 10.1111/j.1468-2427.1994.tb00288.x.

Nagabhushanam, N. (1997) *The Changing Environmental Scenario of a Temple City*. Unpublished Paper, S.V. University, Tirupati, India.

Nakanishi, Y. (2018) Shintoism and travel in Japan. In: Butler, R. and Suntikul, W. (eds) *Tourism and Religion: Issues and Implications*. Bristol, Channel View Publications, pp. 68–82.

Narayanan, V. (1997) "One Tree Is Equal to Ten Sons": Hindu responses to the problems of ecology, population, and Consumption. *Journal of the American Academy of Religion* 65(2), 291–332. DOI: 10.1093/jaarel/65.2.291.

Nash, J. (2012) Re-Examining ecological aspects of Vrindavan pilgrimage. In: Manderson, L., Smith, W. and Tomlinson, M. (eds) *Flows of Faith*. Springer, Dordrecht, The Netherlands, pp. 105–121.

Nelson, L.E. (ed.) (1998) *Purifying the Earthly Body of God: Religion and Ecology in Hindu India*. State University of New York Press, Albany, New York.

Nolan, M.L. and Nolan, S. (1989) *Christian Pilgrimage in Modern Western Europe*. University of North Carolina Press, Chapel Hill, North Carolina.

Nolan, M.L. and Nolan, S. (1992) Religious sites as tourism attractions in Europe. *Annals of Tourism Research* 19(1), 68–78. DOI: 10.1016/0160-7383(92)90107-Z.

Oberdick, U. (1995) Functions of Hindu pilgrimage: secular traits in Tirthayatra. In: Dubey, D.P. (ed.) *Pilgrimage Studies: Sacred Places, Sacred Traditions*. Society of Pilgrimage Studies, Allahabad, India, pp. 69–79.

Olsen, D.H. (2003) Heritage, tourism, and the commodification of religion. *Tourism Recreation Research* 28(3), 99–104. DOI: 10.1080/02508281.2003.11081422.

Olsen, D.H. (2006a) Management issues for religious heritage attractions. In: Timothy, D.J. and Olsen, D.H. (eds) *Tourism, Religion and Spiritual Journeys*. Routledge, London and New York, pp. 120–134.

Olsen, D.H. (2006b) Tourism and informal pilgrimage among the Latter-Day Saints. In: Timothy, D.J. and Olsen, D.H. (eds) *Tourism, Religion and Spiritual Journeys*. Routledge, London and New York, pp. pp. 256–.270.

Olsen, D.H. (2010) Pilgrims, tourists and Max Weber's "ideal types". *Annals of Tourism Research* 37(3), 848–851. DOI: 10.1016/j.annals.2010.02.002.

Olsen, D.H. (2013a) A scalar comparison of motivations and expectations of experience within the religious tourism market. *International Journal of Religious Tourism and Pilgrimage* 1(1), 41–61.

Olsen, D.H. (2013b) Touring sacred history: the Latter-day Saints and their historical sites. In: Hunter, J.M. (ed.) *Mormons and American Popular Culture: The Global Influence of an American Phenomenon*. Vol. 2. Praeger Publishers, Santa Barbara, California, pp. 225–242.

Olsen, D.H. (2016) The Church of Jesus Christ of Latter-day Saints, their "three-fold mission," and practical and pastoral theology. *Practical Matters: A Journal of Religious Practices and Practical Theology* 9, 27–51.

Olsen, D.H. (2019a) The symbolism of sacred space. In: Crous-Costa, N., Aulet, S. and Vidal-Casellas, D. (eds) *Interpreting Sacred Stories: Religious Tourism, Pilgrimage and Intercultural Dialogue*. CAB International, Wallingford, UK, pp. 29–42.

Olsen, D.H. (2019b) Best practice and sacred site management: the case of temple square in salt lake city, Utah. In: Griffinsm, M. and Wiltshier, P. (eds) *Managing Religious Tourism*. CAB International, Wallingford, UK, pp. 65–78.

Olsen, D.H. and Timothy, D.J. (2006) Tourism and religious journeys. In: Timothy, D.J. and Olsen, D.H. (eds) *Tourism, Religion and Spiritual Journeys*. Routledge, London and New York, pp. 1–21.

Olsen, D.H. and Timothy, D.J. (2018) Tourism, Salt Lake City, and the cultural heritage of Mormonism. In: Butler, R. and Suntikul, W. (eds) *Tourism and Religion: Issues, Trends and Implications*. Channel View Publications, Bristol, UK, pp. pp. 250–.269.

Parthasarthy, S. (2002) The new age gurus. In: Sundar, P. (ed.) *For God's Sake: Religious Charity and Social Development in India*. Sampradaan Indian Centre for Philanthropy, New Delhi, pp. 1–40.

Petrillo, C.S. (2003) Management of churches and religious sites. In: Fernandes, C., McGettigan, F. and Edwards, J. (eds) *Religious Tourism and Pilgrimage: ATLAS - Special Interest Group 1st Expert Meeting*. Tourism Board of Leiria, Fatima, pp. 71–86.

Picard, D. and Robinson, M. (eds) (2006) *Festivals, Tourism and Social Change: Remaking Worlds*. Channel View Publications, Clevedon, UK.

Pinkney, A.M. and Whalen-Bridge, J. (2018) *Religious Journeys in India: Pilgrims, Tourists, and Travelers*. SUNY Press, New York.

Pohoaţă, I., Socoliuc, O. and Bostan, I. (2013) The economic impact of religious tourism on the North East region of Romania. *European Journal of Science and Theology* 9, 195–204.

Poria, Y., Butler, R. and Airey, D. (2003) Tourism, religion and religiosity: a Holy mess. *Current Issues in Tourism* 6(4), 340–363. DOI: 10.1080/13683500308667960.

Prime, R. (2008) Saving Krishna's forests. Down to earth. Available at: https://www.downtoearth.org.in/coverage/saving-krishnas-forests-21430 (accessed 21 January 2020).

Prorok, C.V. (1997) Becoming a place of pilgrimage: an Eliadean interpretation of the miracle at Ambridge, Pennsylvania. In: Stoddard, R.H. and Morinis, A. (eds) *Sacred Places, Sacred Journeys*. Geoscience publications, Louisiana State University, Baton Rouge, pp. 117–146.

Raj, R. and Morpeth, N.D. (eds) (2007) *Religious Tourism and Pilgrimage Festivals Management: An International Perspective*. CAB International, Wallingford, UK.

Raj, R. and Griffin, K.A. (2015) *Religious Tourism and Pilgrimage Management: An International Perspective*. CAB International, Wallingford, UK.

Rodrigues, S. and McIntosh, A. (2014) Motivations, experiences and perceived impacts of visitation at a Catholic monastery in New Zealand. *Journal of Heritage Tourism* 9(4), 271–284. DOI: 10.1080/1743873X.2014.899602.

Rodríguez, X.A., Martínez-Roget, F. and González-Murias, P. (2018) Length of stay: evidence from Santiago de Compostela. *Annals of Tourism Research* 68, 9–19. DOI: 10.1016/j.annals.2017.11.001.

Ron, A.S. and Feldman, J. (2009) From spots to themed sites – the evolution of the Protestant Holy land. *Journal of Heritage Tourism* 4(3), 201–216. DOI: 10.1080/17438730802504108.

Rösel, J. (1983) Landed endowment and sacred food. *Archives Eurpeennes De Sociologie* 24(1), 44–59.

Rotherham, I.D. (2007) Sustaining tourism infrastructures for religious tourists and pilgrims within the UK. In: Raj, R. and Morpeth, N.D. (eds) *Religious Tourism and Pilgrimage Festivals Management*. CAB International, Wallingford, UK, pp. 64–77.

Ruback, R.B., Pandey, J. and Kohli, N. (2008) Evaluations of a sacred place: Role and religious belief at the *Magh Mela*. *Journal of Environmental Psychology* 28(2), 174–184. DOI: 10.1016/j.jenvp.2007.10.007.

Saraswati, B. (ed.) (1998) *The Cultural Dimension of Ecology*. Indira Gandhi National Centre for the Arts, New Delhi, India.

Schott, S.B. (2010) "Standing where your heroes stood": Using historical tourism to create American and religious identities. *Journal of Mormon History* 36(4), 41–66.

Shackley, M. (1999) Managing the cultural impacts of religious tourism in the Himalayas, Tibet and Nepal. In: Robinson, M. and Boniface, P. (eds) *Tourism and Cultural Conflicts*. CAB International, Wallingford, UK, pp. 95–111.

Shackley, M. (2001) *Managing Sacred Sites: Service Provision and Visitor Experience*. Continuum, London.

Shackley, M. (2004) Tourist consumption of sacred landscapes space, time and vision. *Tourism Recreation Research* 29(1), 67–73. DOI: 10.1080/02508281.2004.11081432.

Shani, A., Rivera, M.A. and Severt, D. (2007) "To bring God's word to all people": The case of a religious theme-site. *Tourism* 55(1), 39–50.

Shinde, K.A. (2007a) Pilgrimage and the environment: challenges in a pilgrimage centre. *Current Issues in Tourism* 10(4), 343–365. DOI: 10.2167/cit259.0.

Shinde, K.A. (2007b) Visiting sacred sites in India: religious tourism or pilgrimage? In: Raj, R. and Morpeth, N.D. (eds) *Religious Tourism and Pilgrimage Festivals Management: An International Perspective*. CAB International, Oxfordshire, UK, pp. 186–199.

Shinde, K.A. (2008) The environment of pilgrimage in the sacred site of Vrindavan, India. Ph.D. Thesis. Melbourne, Australia, Monash University.

Shinde, K.A. (2011) "This is religious environment": Environmental discourses and environmental behaviour in the Hindu pilgrimage site of Vrindavan. *Space and Culture* 14(4), 448–463.

Shinde, K.A. (2012a) Policy, planning, and management for religious tourism in Indian pilgrimage sites. *Journal of Policy Research in Tourism, Leisure and Events* 4(3), 277–301. DOI: 10.1080/19407963.2012.726107.

Shinde, K.A. (2012b) Shifting pilgrim trails and temple-towns in India: problems and prospects. In: Winter, T. and Daley, P. (eds) *Routledge Handbook on Heritage in Asia*. Routledge, London and New York, pp. 328–338.

Shinde, K.A. (2017) Planning for urbanization in religious tourism destinations: insights from Shirdi, India. *Planning Practice & Research* 32(2), 132–151. DOI: 10.1080/02697459.2016.1198197.

Shinde, K.A. (2018) Augmenting pilgrimages: a religious theme Park in Shegaon, Maharashtra. In: Pinkney, A.M. and Whalen-Bridge, J. (eds) *Religious Journeys in India: Pilgrims, Tourists, and Travelers*. University of New York Press, Albany, New York, pp. 65–90.

Shinde, K.A. and Pinkney, A.M. (2013) Shirdi in transition: Guru devotion, urbanisation and regional pluralism in India. *South Asia: Journal of South Asian Studies* 36(4), 554–570. DOI: 10.1080/00856401.2012.726605.

Shoval, N. (2000) Commodification and theming of the sacred: Changing patterns of tourist consumption in the "Holy Land". In: Gottdiener, M. (ed.) *New Forms of Consumption: Consumers, Culture, and Commodification*. Rowman and Littlefield, Lanham, Maryland, pp. 251–262.

Singh, R.P.B. (1997) Sacred space and pilgrimage in Hindu Society: the case of Varanasi. In: Stoddard, R.H. and Morinis, A. (eds) *Sacred Places, Sacred Spaces: The Geography of Pilgrimages*. Geoscience Publications, Louisiana State University Press, Baton Rouge, pp. 191–207.

Singh, S. (2002) Managing the impacts of tourist and pilgrim mobility in the Indian Himalayas. *Revue De Geographie Alpine* 90(1), 25–35.

Singh, S. (2005) Secular pilgrimages and sacred tourism in the Indian Himalayas. *GeoJournal* 64(3), 215–223. DOI: 10.1007/s10708-005-5649-8.

Smith, V. (1992) The quest in guest. *Annals of Tourism Research* 19, 1–17.

Sofield, T.H.B. and Brent, M. (2001) Sustainability and pilgrimage tourism in the Kathmandu Valley of Nepal. In: Smith, V. (ed.) *Hosts and Guests Revisited: Tourism Issues of the 21st Century*. Cognizant Communication Corp, New York, pp. 257–272.

Srikrishna, S. (1994) Urban air quality management: a case study of Tirupati. In: Singh, H.H., Dube, K.K., Singh, S.B., Kumra, V.K. and Singh, J. (eds) *Patterns of Urban Change in India*. Indrasini Devi Publications, Varanasi, pp. 255–260.

Stausberg, M. (2012) *Religion and Tourism: Crossroads, Destinations and Encounters*. Routledge, London and New York.

Stoddard, R.H. and Morinis, A. (eds) (1997) *Sacred Places, Sacred Journeys*. Geoscience Publications, Louisiana State University, Baton Rouge, Louisiana.

Sullivan, B.M. (1998) Theology and ecology at the birthplace of Krsna. In: Nelson, L.E. (ed.) *Purifying the Earthly Body of God: Religion and Ecology in Hindu*. State University of New York Press, Albany, New York, pp. 247–268.

Swatos Jr., W.H. and Tomasi, L. (eds) (2002) *From Medieval Pilgrimage To Religious Tourism: The Social and Cultural Economics of Piety*. Praeger, Westport, Connecticut.

Tanner, R. and Mitchell, C. (2002) *Religion and the Environment*. Palgrave, New York.

Terzidou, M., Stylidis, D. and Szivas, E.M. (2008) Residents' perceptions of religious tourism and its socio-economic impacts on the Island of Tinos. *Tourism and Hospitality Planning & Development* 5(2), 113–129. DOI: 10.1080/14790530802252784.

Timothy, D.J. (2011) *Cultural Heritage and Tourism: An Introduction*. Channel View Publications, Clevedon, UK.

Timothy, D. (2012) Religious views of the environment: Sanctification of nature and implications for tourism. In: Holden, A. and Fennell, D.A. (eds) *The Routledge Handbook of Tourism and the Environment*. Routledge, London and New York, pp. 53–64.

Timothy, D.J. and Iverson, T. (2006) Tourism and Islam: considerations of culture and duty. In: Timothy, D.J. and Olsen, D.H. (eds) *Tourism, Religion and Spiritual Journeys*. Routledge, London and New York, pp. 186–205.

Timothy, D.J. and Olsen, D.H. (eds) (2006) *Tourism, Religion and Spiritual Journeys*. Routledge, London and New York.

Tomasi, L. (2002) Homo viator: from pilgrimage to religious tourism via the journey. In: Swatos Jr., W.H. and Tomasi, L. (eds) *From Medieval Pilgrimage To Religious Tourism: The Social and Cultural Economics of Piety*. Praeger, Westport, Connecticut, pp. 1–24.

Trivedi, S. and Agrawal, M. (2003) Air quality status in Varanasi city. *Journal of Environmental Studies and Policy* 6(2), 75–85.

Tuan, Y.-F. (1974) *Topophilia: A Study of Environmental Perception, Attitudes and Values*. Prentice Hall Inc, New Jersey.

Turnbull, C. (1981) Holy places and people of India. *Natural History* 90(9), 76–81.

Turner, V. and Turner, E. (1978) *Image and Pilgrimage in Christian Culture*. Columbia University Press, New York.

van Der Veer, P. (1988) *Gods on Earth: The Management of Religious Experience and Identity in a North Indian Pilgrimage Centre*. The Athlone Press, London.

Varady, R.G. (1989) Land use and environmental change in the Gangetic plain: nineteenth century human activity in the Banaras region. In: Freitag, S.B. (ed.) *Culture and Power in Banaras: Community, Performance, and Environment*. University of California Press, Berkeley, California, pp. 229–245.

Verschuuren, B., Wild, R., McNeely, J.A. and Oviedo, G. (2010) *Sacred Natural Sites: Conserving Nature & Culture*. Earthscan, London.

Vidyarthi, L.P., Jha, M. and Saraswati, B. (1979) *The Sacred Complex of Kashi: A Microcosm of Indian Civilisation*. Concept Publishers, New Delhi, India.

Voye, L. (2002) Popular religion and pilgrimages in Western Europe. In: Swatos, W.H. and Tomasi, L. (eds) *From Medieval Pilgrimage To Religious Tourism: The Social and Cultural Economics of Piety*. Praeger, Westport, Connecticut, pp. 115–135.

Vukonić, B. (1996) *Tourism and Religion*. Pergamon, New York.

Wong, C.U.I. (2018) The monks and nuns of Pu-Tuo as custodians of their sacred Buddhist site. In: Butler, R. and Suntikul, W. (eds) *Tourism and Religion: Issues and Implications*. Channel View Publications, Bristol, UK.

Wong, C.U.I. and Ryan, C. (2013) Tourism at Buddhist shrines on Putuoshan: a co-existence of the secular and profane. In: Ryan, C. and Huang, S. (eds) *Tourism Tourism in China: Destinations, Planning and Experiences*. Channel View Publications, Bristol, pp. 250–261.

Wong, C.U.I., McIntosh, A. and Ryan, C. (2013) Buddhism and tourism: perceptions of the monastic community at Pu-Tuo-Shan, China. *Annals of Tourism Research* 40, 213–234.

Wong, C.U.I., McIntosh, A. and Ryan, C. (2016) Visitor management at a Buddhist sacred site. *Journal of Travel Research* 55(5), 675–687. DOI: 10.1177/0047287514563164.
World Tourism Organization (2011) *Religious Tourism in Asia and the Pacific*. UNWTO, Madrid.
YES Bank (2014) *Towards Sustainable Spiritualism: Buddhist Tourism - Linking Cultures, Creating Livelihoods*. YES Bank, New Delhi.
Young, M. (1999) The social construction of tourist places. *Australian Geographer* 30(3), 373–389. DOI: 10.1080/00049189993648.
Zamani-Farahani, H. and Henderson, J.C. (2010) Islamic tourism and managing tourism development in Islamic societies: the cases of Iran and Saudi Arabia. *International Journal of Tourism Research* 12, 79–89.

2 Pilgrimage, Religious Tourism, Biodiversity, and Natural Sacred Sites

Daniel H. Olsen*
Brigham Young University, Provo, Utah, United States

Introduction

For millennia, people have travelled to sites deemed sacred in search of transcendental, epiphanic, or transformative experiences to find meaning and understanding regarding the nature of human existence – experiences that must be sought out beyond the geographies of their daily lives (Wearing *et al.*, 2016; Olsen, 2017). Mountains, caves, volcanoes, rivers, lakes, springs, and other natural phenomena are some of the oldest venerated sites in the world (Verschuuren *et al.*, 2010c). Because of the importance of the natural environment for human survival, early humans travelled to these natural sites to engage in prehistoric and pagan rites and worship ceremonies to gain favour from the gods or from the earth itself in the face of a hostile world, with some of these sites dating as far back as 10,000 years (Timothy, 2011). Travel to these natural sacred sites marked the beginning of the development of both spiritual and religious traditions around the world. Even today, First Nations people in North America, and tribal religions in Africa, Hinduism, and Shintoism incorporate naturalistic veneration and animistic beliefs into their religious traditions (Jones and Pennick, 1995), as do those who follow the pantheistic ecosophy of the green movement within the New Age movement (Timothy and Conover, 2006; Gábrišová, 2014) or incorporate mid-eighteenth century romanticism into deep ecological and environmentalist movements (Sessions, 1987; O'Neil, 2001; Ferber, 2010; Gottlieb, 2010; Bremer, forthcoming).

Travel to natural sacred sites (also referred to in the literature as 'sacred natural sites') for worship purposes remains a significant motivation for people to travel. Some of the more prominent examples of the thousands of natural sacred sites that presently exist in the world today, prompting pilgrimage and religious tourism travel, include Uluru (Ayers Rock) in Australia, Mt Sinai in Egypt, the Ganges and Brahmaputra rivers in India and Pakistan respectively, the Batu Caves in Malaysia, Mt Olympus in Greece, Mt Kailash in western Tibet, Mt Fuji in Japan, and the monolithic stones near Meteora, Greece. Many human-built sacred sites have also been established on or within sacred natural landscapes, such as holy wells on top of sacred springs (e.g. Ireland), monasteries built at the foot of sacred mountains (e.g. Saint Catherine's Monastery, Egypt, Mt Kōya, Japan, and Shinto temples found at the base of Mt Fuji), and temples along holy rivers (e.g. Kashi Vishwanath Temple, Varanasi). In addition, aspects of the natural environment have long influenced the way in which religious groups have developed the architecture and ideology behind their built

*dholsen@byu.edu

sacred sites. Early medieval cloisters and monasteries, for example, were designed to be cosmologically akin to sacred mountains, with their gardens symbolizing the Garden of Eden from the Book of Genesis (Helms, 2002). Another example comes from Garcia-Fuentes's work in Chapter 8 in this volume where Mt Monserrat in Spain was used as the archetypical model for the building of the Basílica de la Sagrada Família in Barcelona.

However, with technological advances in transportation and telecommunications since World War II, religious sites, including natural sacred sites, have seen an unprecedented growth of visitation. While travel to these sites under the guise of ecotourism can be a tool to help maintain natural sacred sites (Papayannis and Mallarach, 2008; Nyamweru and Kimaru, 2008), increasing visitation has led to several concerns among scholars, environmentalists, and indigenous leaders regarding the long-term sustainability, preservation, and conservation of these sites (e.g. Verschuuren *et al.*, 2010b). This has become a particular concern at international levels in recent years, where sacred sites are increasingly becoming a topic of discussion within international legal frameworks on the preservation of global biodiversity (Gomez *et al.*, 2010). However, emphasis in the academic literature related to the impacts of pilgrimage and religious tourism still tends to focus on the human-built environment, neglecting to consider in large part the impacts of pilgrimage and religious tourism development in natural sacred settings.

Therefore, following Timothy (2012), the purpose of this chapter is to examine the impacts of pilgrimage and religious tourists on natural sacred sites. After discussing the views of the environment held by various religious groups and the tensions between secular and religious modes of environmentalism, definitions and different types of natural sacred sites are considered. Attention is then turned to examine how pilgrimage and religious tourism affect natural sacred sites, followed by a discussion of the management of sacred natural sites and their role in the preservation of biodiversity, before concluding.

Religious Views of the Environment

Religion and the environment are strongly intertwined, with several holy texts, such as the Judaeo-Christian Bible, the Quran, the Sri Guru Granth Sahib (SGGS), and the Bhagavad Gita making statements regarding the creation of the world and the relationship between God, humanity, and nature. For example, in Genesis 1:28 it says: 'And God blessed them, and God said unto them, Be fruitful, and multiply, and replenish the earth, and subdue it: and have dominion over the fish of the sea, and over the fowl of the air, and over every living thing that moveth upon the earth.' The Quran states that all of Allah's creations glorify God (Quran 22:18) and calls upon people to recognize the damage they do to the environment: 'Corruption doth appear on land and sea because of (the evil) which men's hands have done, that He may make them taste a part of that which they have done, in order that they may return' (Quran 30:41). In Hinduism, the Bhagavad Gita highlights the interconnectedness of humans and nature:

> Know that Prakriti, Nature, and Purusha, Spirit, are both without beginning, and that temporal changes and Gunas, conditions, come all from nature.
>
> Nature is the source of all material things: the maker, the means of making, and the things made. Spirit is the source of all consciousness which feels pleasure and feels pain.
>
> The spirit of man when in nature feels the ever-changing conditions of nature. When he binds himself to things ever-changing, a good or evil fate whirls him round through life-in-death.
>
> But the Spirit Supreme in man is beyond fate. He watches, gives blessing, bears all, feels all. He is called the Lord Supreme and the Supreme Soul.
>
> He who knows in truth this Spirit and knows nature with its changing conditions, wherever this man may be he is no more whirled round by fate (13:19–23).

The Sri Guru Granth Sahib, the Sikh holy scripture, is also replete with scriptural injunctions regarding the relationship between humanity and nature. For example:

> Air is our Guru, water our father, and the great earth our mother.

> Day and night are the female and male nurses in whose laps the whole universe plays.
>
> Good and bad deeds are all disclosed in the presence of Righteousness.
>
> Our actions take us near or far.
>
> Those who remember the Name earn true success.
>
> Nanak says their faces shine, and they take many with them to liberation.
>
> (SGGS, M1, p. 8; see Singh, 1995, p. 67)

However, even though many modern environmental ethics and movements are rooted in religious beliefs and values, some of the religious doctrines underlying these environmental efforts have come under considerable scrutiny, with several scholars debating whether this religious influence has been positive or negative (e.g. Guth *et al.*, 1995; Woodrum and Wolkomir, 1997; Biel and Nilsson, 2005). Much of this consternation arose with White's (1967) view that Judaeo-Christian beliefs regarding the environment were the cause of present-day environmental problems. According to White, the biblical injunction to 'have dominion' over nature has led to an anthropocentric environmental tradition which seems to promote wanton destruction of the non-human environment. As Timothy (2012) notes, fundamental Christians seem to promote an anti-conservation or anti-green ethos. Islam, Hinduism, and Sikhism, however, have not been immune from these types of criticisms. Even though these faiths, especially Hinduism and Sikhism, are viewed as 'ecosophical' because of their focus on going 'beyond anthropocentrism towards holistic consciousness of the universe' (Lourdunathan, 2012, p. 167), what is preached and what is practiced are two different things. As Holden (2003, pp. 98–99) notes, while 'religions in non-Western societies tend to have creed which discourage a demarcation between human and non-human environment [,] ... industrial development has resulted in ... encourag[ing] a different perception of the environment to the one commonly held in the West'. For example, as noted in Chapter 11, participants in the Hajj demonstrate a lack of awareness or concern regarding their impacts on the natural environment.

However, much of the modern environmental movement has been rooted in religious belief and value systems, whether based on the spiritual nature of John Muir's spiritually-based views of nature transcending humanity (Powici, 2004; Timothy, 2012), environmental eschatology (Guth *et al.*, 1995), or the move from environmental 'domination' to the environmental 'stewardship' ethic within Judaeo-Christian thought (Ehrenfeld and Bentley, 1985; Curry-Roper, 1990; Shibley and Wiggins, 1997; Woodrum and Wolkomir, 1997). Some scholars who argue that modern environmentalism should be based on purely secular lines (see Dunlap, 2006; Harper, 2008; Berry, 2013) question whether the influence of religion on this movement has been positive or negative (Vogel, 2001), particularly when it comes to seemingly incongruent religious teachings with present-day environmental views and activism. However, others note that religious leaders and social commentators from many faiths teach against the environmental excesses of an increasingly consumerist-oriented society (e.g. De Silva, 1998; Wei, 2004; Jenkins, 2013). In addition, there are several faith-based non-government religious/environmental organizations (Ellingson *et al.*, 2012), as well as larger religious interfaith organizations, that work towards environmental sustainability, ecojustice, and environmental reform at different scales in many areas of the world. These organizations include the National Religious Partnership for the Environment, the National Council of Churches (USA), the World Council of Churches, and the Parliaments of World Religions, among several others (Tucker and Grim, 2001; Gardner, 2003; Smith and Pulver, 2009; Bistcotti and Biggart, 2014).

Natural Sacred Sites

According to Verschuuren *et al.* (2017), one of the earliest attempts of conservationists to create a working definition for natural sacred sites was Oviedo and Jeanrenaud (2007, p. 77), who defined natural sacred sites as 'natural areas recognized as sacred by indigenous and traditional peoples, as well as natural areas recognized by institutionalized religions or faiths as places for worship and remembrance'. Another more commonly accepted definition of natural sacred sites is: 'areas of land or water having special spiritual significance to people and communities' (Wild and McLeod, 2008, p. xi). Aitpaeva (2013, p. 7) expands on this definition to include the performance of 'sacral practices' at these sites. The fact that in many instances religious or spiritual rituals or practices are performed at these natural sacred sites implies that they are 'in some way holy, venerated or consecrated and so connected with religion or belief systems, or set aside for a spiritual purpose' (Verschuuren *et al.*, 2010c, p. 2).

Whether venerated by mainstream faiths, spiritual organizations and individuals, or local indigenous populations, natural sacred sites exist all around the world. They occur at different scales, ranging from single natural elements, such as an individual tree or rock, to entire natural landscapes (Verschuuren *et al.*, 2010c), and, at the fullest extent, the entire world. In the case of larger landscape areas, it may be difficult to determine the limit of a natural sacred space unless it is firmly demarcated in some way. This is partly why there is no exhaustive inventory listing of these sites. However, Dudley *et al.* (2010) suggest that there are certain regions of the world where there seems to be a concentration of these types of sites, including India, where there are an estimated 100,000–150,000 natural sacred sites; Africa, where there are 920 in Tanzania and 2,000 in Ghana alone (Ntiamoa-Baidu, 1995); and in China, particularly in the Xishuangbanna, South Yunnan, where there are several sacred hills (see Shengji and Wild, 2010). More broadly Rutte (2011) suggests that 250,000 natural sacred sites globally is probably an accurate number.

Natural sacred sites are an integral part of indigenous religions and religions, such as Hinduism and Buddhism, that are, as noted above, more 'ecosophical', with a greater emphasis on attachment to natural landscapes (Mazumdar and Mazumdar, 1993). Natural sacred sites are also important pilgrimage destinations for members of the New Age movement (Ivakhiv, 2003). This movement capitalizes on modern shifts towards individualization and alternate spiritualties and sacred spaces through a focus on 'self-improvement for mind, body, and soul through harmony [with] the universe, ancient spiritual traditions, the sanctity of nature, earth's powers, animism, extraterrestrial visitation, past-life regression, yoga and meditation, and self-deification' (Timothy, 2012, p. 37; see also Timothy and Conover, 2006). To do this, elements of different religious traditions, including natural sacred sites, are co-opted and commodified to cater to the demands of the new and ever-expanding spiritual marketplace (York, 2001; Redden, 2005; Muir, 2007; Gauthier *et al.*, 2013; Olsen, 2019). Neo-pagans are also a group that uses natural sacred sites for worship purposes, travelling to archaeological and natural heritage sites to worship pre-Christian religions or deities (Rountree, 2006).

There are several types of natural sacred site that are frequented by religious, indigenous, and new age/neo-pagan groups.

Sacred mountains and volcanoes

Mountains and volcanoes have long been venerated by religious and cultural groups, many of which were founded on or in the shadow of these geomorphic features. Mountains and volcanoes serve as the abode of the gods (Bernbaum, 2012), and in some cases were viewed as the physical embodiment of gods (Timothy, 2012). Like other aspects of sacred geography, sacred mountains and volcanoes also serve as providers of a sense of psychological well-being for cultural groups, such as the Navajo in the USA, where group identity is intimately tied to the natural landscape (Griffin-Pierce, 1997), or in cases where general leisure-based tourists use these landscapes for therapeutic purposes (Heintzman, 2009).

Bernbaum (2012) suggests that there are three ways in which mountains and volcanoes gain an aura of the sacred. First, they may be singled out by indigenous and mainstream religious traditions as being the focus of myth and ritual practice. Archaeological ruins on the summits of many mountains in the Andes seem to suggest that prehispanic cultures venerated these mountains that were believed to be the abode of the gods, which gods were the source of water and manipulators of weather patterns (Reinhard, 1985). Mountains in ancient

Chinese culture were viewed as pillars that separated heaven and earth (Timothy, 2012). The same can be said of volcanoes, which are viewed as guardians of the people (Westervelt, 1916). In terms of specific mountains and volcanoes, Mt Kailash in Tibet is viewed by Hindus as the centre of the universe; the San Francisco Peaks in the USA are considered by the Hopi people as the home of ancestral spirits (Bernbaum, 2010); Mt Bromo in Indonesia is held as sacred to the Tenggerese people in Eastern Java (Choe and Hitchcock, 2018); and the Kilauea volcano in Hawaii is viewed as the embodiment of the volcano goddess Pele (Bernbaum, 2006).

Second, mountains and volcanoes are marked as sacred by the presence of temples, stones, grottos, and holy people. As noted above, many monasteries have been built at the foot of sacred mountains. Pura Besakih, the most important Hindu temple on Bali, is located high on the slopes of Mt Agung volcano (Stuart-Fox, 2002), and the Taktsang Palphug Buddhist Monastery (also known as the 'Tiger's Nest') is perched on a mountain cliff 900 metres above the Paro Valley Floor in Bhutan (Pommaret, 2009). There are also several ruined temple complexes and statues throughout the mountains and volcanoes of the island of Java (Kaehlig *et al.*, 2012). Third, mountains and volcanoes may be part of a broader evocative and awe-inspiring natural landscape that sets them apart as being special. As Bernbaum (2006, p. 304) writes: '... mountains have the power to awaken a sense of the sacred. Their soaring summits, the clouds and thunder that swirl around their peaks, the life-giving waters that flow from their heights, these and other characteristics imbue them with an aura of mystery and sanctity' (see also Chapter 4, this volume).

Caves and grottoes

Caves and grottoes are the opposite of mountains and volcanoes, in that they are hidden from sight and in many cases lacking natural light (Kiernan, 2015; Sponsel, 2015). Visits to caves and grottoes have been described as 'crossing into the subterranean world' and 'characterized as liminal, a transitional state from the place of ordinary above ground activities into a place of extraordinary underground activities ... [Caves] become ... a gateway into the spiritual dimension of human experience, or even into the supernatural world' (Sponsel, 2015, p. 506). In many faiths, caves and grottoes represent birth, rebirth, and regeneration (Timothy, 2012). Caves and grottoes are also repositories of art and places of burial (Sponsel, 2015). This is the case in Christianity, where it is believed that Jesus was born in a grotto (Bacci *et al.*, 2012). Holy men and women throughout history have also secluded themselves in caves and grottoes, with Buddhist monks and nuns using caves as homes and meditation spaces in northern India as early as 2,500 years ago (Sponsel and Natadecha-Sponsel, 2004). Daoists believe that sacred caves contain purified forms of *qi*, or the life force that animates the world, and as such meditate there (Kiernan, 2015). Several Christian churches and chapels and Buddhist monasteries are built inside caves, such as in Italy where there are several grottoes that have been transformed into churches (Robinson, 1930; Cennamo *et al.*, 2016). The Magao Caves in the Gangsu province in China house 492 Buddhist temples (Whitfield *et al.*, 2000), and the Cave of Hira where Muhammad is believed to have been visited by the angel Gabriel is one of Islam's holiest places (Timothy, 2012).

Sacred water

Water has long been held in reverence by people around the world because of its importance to humankind's survival (Altman, 2002). This is particularly the case in places that suffer from a limited water supply. Many divinities within religious traditions are related to water, such as Poseidon (Greek god of the sea) and Nun (Egyptian god of the sacred ocean). Water is also an important component to many creation myths (Leeming, 2010), religious ceremonies and rituals, healing (Timothy, 2012), and, like caves, are symbols of birth and rebirth (Altman, 2002).

Sacred water revolves around several topographical features related to water, such as lakes, rivers, and springs (Altman, 2002), and can even affect other natural elements, such as sacred water use for irrigation leading to the development of sacred vegetation (Pandy, 2000). One of the great examples of a holy river is the Ganges River in India. While large rivers and are deemed sacred in Hinduism, the Ganges is held to be especially sacred, as it is believed to be the physical embodiment of the goddess Ganga (Agoramoorthy, 2015). Annually, millions of people come to Varanasi – a *tirtha* or place of pilgrimage near a river – to bathe themselves in the river for healing purposes, to be cleansed from sin, and for salvific purposes (Flood, 1966; Alley, 2002; Agoramoorthy, 2015). To facilitate these religious rituals, a sacred environment comprised of temples and shrines has been built, and religious specialists are available to help pilgrims with their sacred and death-related rites (Gesler and Pierce, 2000; Sharma, 2020).

Holy wells, as 'repository[ies] of spiritual power' (Rattue, 1995, p. 21) and the home of spirits, are found in many faith traditions. Generally tied to the feminine, holy wells are also sites of healing, magic, and entrances to otherworldly wisdom (Varner, 2009; Foley, 2011). Holy wells are most often associated with Ireland, where there over 3,000 wells tied to Catholicism (Logan, 1980; Varner, 2009). The water temples of Bali are important locations where both the worship of water deities and decisions regarding irrigation flows and water usage take place (Zurich, 2002). People still visit and ritually bathe in these waters, even when water sources are heavily polluted due to industrial and human effluent (Semwal and Akolkar, 2006; Lüthi, 2010; Colopy, 2012; Das and Tamminga, 2012).

Sacred trees and groves

Sacred groves and trees exist all over the world and have been used for religious and livelihood purposes for thousands of years (Barrow *et al.*, 2010). Trees and groves have been tied to the home of spirits, the burial of chieftains and spiritual leaders, marriage practices, and are at times adorned or marked with items deemed sacred by communities that have a deep respect for the environment (Bhagwat and Rutte, 2006; Barrow *et al.*, 2010; Ormsby, 2011). Sacred groves are often protected by religious taboos, such as cutting down trees and killing animals within their boundaries (Hughes and Chandran, 1998). Because of these taboos, sacred groves are also usually uninhabited (Barrow *et al.*, 2010), and as such are considered one of the earliest attempts by humans to engage in nature conservation (Kiernan, 2015). Many sacred groves protect endangered tree species, conserve old tree growth, and serve as corridors for animal migrations, particularly in densely-developed human landscapes (Bhagwat and Rutte, 2006; Salick *et al.*, 2007). Sacred groves also vary in size and type of governance, ranging from national governments under the guise of protected areas to communities with little government interference (Ormsby, 2011). In India, there are over 4,000 sacred groves (Khan *et al.*, 2008), and in Japan, sacred groves are associated with Shinto temples (Albin and Berwick, 2004; Pungetti, 2013). Some of the more prominent sacred groves include the Lumbini Grove in the Rupandehi District of Nepal (Weise, 2013) and the Osun-Osogbo Sacred Grove in Nigeria (Woosnam *et al.*, 2016).

Sacred plants and animals

In addition to trees, many of the world religions hold certain plants and animals as sacred, meaning that they are either exclusively dedicated to a god or some religious purpose (e.g. ritual sacrifice) or are safeguarded by religious tradition (McIvor and Pungetti, 2012). These species are believed to embody religious, spiritual, cultural, and biological values and ideals (Pungetti, 2012), and are usually significant species within an ecosystem (Ramakrishnan and Harmonand, 2003). As such, sacred sites and religious groups act as protectors of biological diversity and endemic and endangered species. For example, sacred caves in Madagascar and Thailand act as safe havens for endemic bat species (Sponsel and Natadecha-Sponsel, 2004; Fernández-Llamazares *et al.*, 2018). Religious standards, values, and taboos towards the environment have also led religious communities to be proactive in preserving endangered wildlife. Examples include pheasants in Shichuan, China (Nan *et al.*, 2012); lizards in New Zealand (Roberts, 2012); salmon and leatherback sea turtles off the western coast of the USA (Douros, 2012); eels and crocodiles in Madagascar (Andriamarovololona and Jones, 2012); the bowhead whale in north-west Alaska (Lowenstein, 1993); the long-tailed macaque monkey in Bali (Wheatley, 1999); and bald eagles in the USA (McIvor and Pungetti, 2012). Plants that are sacred and are protected by religious communities and sacred sites include the lotus flower (*nelumbo nucifera*), which is sacred to Buddhists and Hindus (Ward, 1952); the Mediterranean cypress (*cupressus sempervirens*), which is sacred to multiple religious groups in Turkey (McIvor and Pungetti, 2012); and plants regarded as sacred or magical that contain psychoactive or hallucinogenic agents that lead to intense spiritual experiences (Díaz, 1977; Basset, 2016) or have therapeutic or medicinal properties (King *et al.*, 1997; Varghese *et al.*, 2010King *et al.*, 1997).

The Impacts of Pilgrimage and Religious Tourism on Natural Sacred Sites

Tourism is generally viewed as a system of related industries that are designed to meet the wants and needs of people travelling outside of their home region for leisure or business purposes (Leiper, 1979; Mill and Morrison, 2002). As such, tourism studies have historically tended to be heavily oriented towards 'business prerogatives' (Bianchi, 2009, p. 486) with an emphasis on marketing, promotion, the quantification of tourist motivations and expenditures, and the development of tourism management techniques that are 'of value to the practicalities of the business of tourism' (Sharpley, 2011, p. 12). While concerns regarding the negative economic, sociocultural, and environmental impacts of tourism date back to the 1960s (Saarinen, 2006), the addition of 'sustainability' into tourism discourses has led to calls for more holistic views of tourism and its impacts (e.g. Wall and Mathieson, 2006; Fleming and Toepper, 1990; Archer *et al.*, 2005).

Within the context of the environmental impacts of tourism, many tourism operators and marketers tend to promote neo-classical economic, anthropogenic, and utilitarian perspectives of nature at the expense of long-term, environmentally sound and intrinsic considerations (Hall, 2013). More recently, there have been increasing appeals for a more ecocentric and intrinsic-approach to the tourism development of natural capital, where sentient and non-sentient aspects of the natural environment are recognized as having inherent and moral value and a right 'to be' (e.g. Holden, 2003, 2009; Yerbury and Boyd, 2018). As such, the natural environment should be developed and managed in a way that the self-destructive qualities of tourism that lead to natural sites being 'loved to death' (Croall, 1995; Sharpley, 2009) can be avoided. For example, ecotourism – a sub-niche of tourism that combines a nature-based visit and experience with minimizing the environmental impacts of that visit (Blamey, 1997) – while having its problems, has long been promoted as a more environmentally-friendly and morally less objectionable type of travel style. As Nyamweru and Kimaru (2008) note within the context of sacred sites, ecotourism development of the *kaya* forests in Kenya has led to 'appropriate' cultural tourism, including increasing the income of residents and increased conservation efforts. However, even with these calls for sustainability, different tourism stakeholders take separate ethical approaches towards the use of the natural or non-human environment (Singh and Kaur, 1986; Holden, 2003). For instance, even though nature tourists express higher levels of concern for the environment than other types of tourist, they are influenced by a wide range of intrinsic and instrumental values, which values are sometimes incongruent with their observed behaviour (Winter, 2007).

Pilgrimage and religious tourism is sometimes viewed as a form of 'soft' tourism that involves more environmentally mindful travel (Lusby, 2017). Within these travel styles (Adler, 1989), the use of natural sacred sites for pilgrimage and religious tourism purposes can have positive environmental impacts. For example, tourism development of natural sacred sites can lead to both the preservation of and the transmission of the sacred values inherent within these sites to visitors (Goral, 2011; Chadha and Onkar, 2016). This is particularly acute when the natural landscape is an essential element for religious and/or spiritual growth, transformation, and well-being (e.g. Heintzman, 2009; Sharpley and Jepson, 2011). The use of 'organic' and the creation of 'purposeful' religious trails (Timothy and Boyd, 2015) for pilgrimage and religious tourism purposes, while leading to certain environments being 'hardened' or permanently changed by religious tourism development, helps to preserve other natural areas long-term (Trombino and Trono, 2018). Pilgrimage and religious tourism also encourage 'slow tourism', which leads to positive encounters with human-built and natural environments and helps to uphold ecological community values (Howard, 2012; Singh, 2012; Kato and Progano, 2017).[1]

However, travel to and participation in rituals in natural settings can lead to several environmental, and hence sociocultural, problems and concerns (see Chapter 1). For example, some Native American communities are upset by the environmental effects of tourism on Mount Shasta in California, which effects include trampling, erosion, litter, overcrowding, and the lack of respect tourists show for the sacred site. This is particularly the case with New Age tourists who camp on the mountain, whose behaviour causes recreational users to complain 'of nudity, drumming, or chanting in the meadow areas and of the construction of altars and medicine wheels, saying that they detract from their nature experience' (Huntsinger and Fernández-Giménez, 2000, p. 540). Mass pilgrimage and religious tourism can also lead to increased deforestation, damage to fauna, air and water pollution, and solid waste, all of which strains local infrastructure (Shinde, 2010, 2012a). This is exacerbated in instances where religious sites are located inside protected areas (Seshardi and Ganesh, 2011). Human-built sacred sites, many of which are found within natural sacred sites, are also impacted by mass pilgrimage and religious tourism.

As such, environmental issues caused by pilgrims and religious tourists, as noted above, can be seen as hypocritical, particularly when a faith's theological considerations regarding the importance of taking care of the natural environment is not shown through the actions of pilgrims and religious tourists (Huntsinger and Fernández-Giménez, 2000). As Shinde (2018) notes, during the month-long 201 km pilgrimage procession from Alandi and Dehu to Pandharpur, India, over a million pilgrims leave heavy environmental burdens on the towns in which the *'palkhi'* or the procession stops, as well as the natural spaces between the towns in the form of garbage and human waste (i.e. open defecation). This is also seen in Chapter 8, where *Hajj* attendees are at best ambivalent towards their macro- and micro-environmental impacts (see El Hanandeh, 2013). Increasing numbers of pilgrims and religious tourists, particularly at certain peak times related to religious festivals and processions, can also intensify environmental impacts related to overpopulation and urbanization (Shinde, 2007).

Governance and the Preservation of Biodiversity

In addition to being socially constructed places of pilgrimage and tourism consumption, natural sacred sites serve as the both tangible and intangible systems and reservoirs of knowledge and history of local, national, and international significance. Indeed, religious beliefs and practices have in many instances led to the preservation of biodiversity with sacred natural landscapes. In one study (Anderson *et al.*, 2005) found that sacred Buddhist sites around Mount Kawa Karpo in Tibet were more ecologically and ethnobotanically unique than sampled non-sacred sites. In Ethiopia, where most of the forests have been sacrificed for agricultural purposes, Orthodox churches and their forested lands are instrumental in the preservation of the little biodiversity that is left (Abbott, 2019), and in Italy, Roman Catholic religious sites, many of which are located in natural areas, have also been seen as reservoirs of biodiversity (Frascaroli, 2013). The same can be said of the northern districts of the state of Kerala, India, where sacred groves preserve diverse bird life in the face of expanding urbanization (John, 2019). This is why Chan *et al.* (2019) refer to natural sacred sites as 'cultural commons'. Natural sacred sites also function – and have functioned historically – as areas of biodiversity conservation as well as the site of material resources and biodiversity for local communities (Samakov and Berkes, 2017; Gadgil, 2018). In fact, in many indigenous communities it is difficult to partition natural sacred sites from the natural resources needed for economic survival. Because of their importance within the context of economic practices, McWilliam (2009) and Samakov and Berkes (2017) further refer to natural sacred sites as part of a more generalized 'spiritual commons', which include the ideas, practices, and collective norms and values that form 'social practice and collective endeavour[s]' (McWilliam, 2009, p. 164) within these sites.

Samakov and Berkes (2017, p. 424) argue that even though natural sacred sites are an important cultural, physical, and spiritual resource, these sites do not always offer 'intrinsic or instrumental value'. Rather, these sites should be considered in terms of their 'relational value' (see Chan *et al.*, 2016), in that 'these sites have values that derive from the relationship that people have to a particular site' (Samakov and Berkes, 2017, p. 424). In the case of natural sacred sites, the conservation and use of these sites revolve around the moral and practical decisions stakeholders make regarding whether these sites should be maintained because of their material (read: economic/monetary) value or because it is a moral imperative that they continue in perpetuity (Bhagwat, 2009). These decisions are also based on issues relating to who has the authority to demarcate a natural site as sacred and control the uses of that site. Indeed, because of their intrinsic and instrumental value, these sites 'always have institutions and rules associated with them' (Verschuuren *et al.*, 2010c, p. 2). Ndoro and Wijesuriya (2015, p. 135) demonstrate that Africa and Asia have a long history of pre-colonial 'traditional' heritage management practices related to natural sacred sites – that '[t]he influence of religion [has] led to well-established principles, procedures, and practices in diverse forms of conservation, management, and long-term care for heritage' in these world regions. In this vein, Matose *et al.* (2019, p. 335) note that in Africa, the pre-colonial demarcation of natural sacred sites in the region was based on the exercising of power by dominant cultural groups over resources rather than a 'reflect[ion of] egalitarian and consensual conservation'.

These institutions and rules relating to the management of natural sacred sites can in some cases lead to contestations between different cultural groups. For example, Wang (2019) notes that Tibetan Buddhists consider mountains as 'mountain-persons' or as living sacred entities, and as such offer articles of clothing to the mountains. However, the Han Chinese in the area see these offerings as litter, which has raised concerns over waste-management practices in the region. Blake (2001) also notes the conflict between Navajo peoples and the Bureau of Land Management/US Forest Service regarding the ways in which their six sacred mountains which are located on federal lands are used for recreational and resource extractive activities without their consultation (see Hooker, 1994; Blatt, 2005; Carpenter, 2006). This has led Ndoro and Wijesuriya (2015) to suggest that the sacred nature of natural sites leads to different sets of values and conservations challenges, particularly in the context of these sites acting as cultural and spiritual commons. As Samakov and Berkes (2017, pp. 424–425) argue, that natural sacred sites cannot be owned exclusively by one group because of the relational value they have to multiple stakeholders. Therefore, these sites need to be given 'public property status to be open to all potential visitors who may have relational values concerning that site'.

There are also contestations between indigenous groups and governments at different scales regarding the ownership and management of sacred sites. Indigenous groups around the world have strong sacred connections to their natural sacred sites, viewing these locations as 'ecologies that are alive with the enduring processes of creation itself' (Battiste and Henderson, 2000, p. 9). As such, these 'land-based' religious groups (Kuokkanen, 2006) have a difficult time reducing the term 'sacred' to western European colonial concepts such as 'concepts conveyed by English words such as "socialconstruction", "essentialism", "representation", or the positivist categories and definitions of human legislation' (Coleman, 2016, p. 65). With European colonialism, indigenous communities were disenfranchised and removed from their sacred geographies within differing colonial ontologies and productions of space. Even though many post-colonial efforts seek to return the rights of ownership of sacred sites to indigenous populations, many indigenous natural sacred sites can be considered 'diasporic space' for those whose lands have not been returned (Coleman, 2016), particularly in instances where indigenous natural sacred sites are found within protected natural areas (Verschuuren *et al.*, 2010a).

Within this context, not only have national governments begun to work with indigenous groups to either co-manage or return these lands to indigenous communities – albeit with mixed results (e.g. Lane *et al.*, 2003; Hooker, 1994; Schlosberg and Carruthers, 2010; Liljeblad and Verschuuren, 2019) – but also with other international organizations and working groups. The International Indigenous Forum on Biodiversity (1998), for example, has argued that indigenous knowledge systems need to be incorporated in the creation of viable sustainable management policies and practices related to sites of biodiversity. The World Bank report *The Role of Indigenous Peoples in Biodiversity Conservation* (Sobrevila, 2008) also strongly recommends that indigenous people need to be actively engaged in the decision-making processes related to these sites (Stewart-Harawira, 2012). The contested nature of land ownership and management, however, has led to intervention at the international level by organizations and groups that work hard to find holistic ways to maintain the integrity of natural sacred sites and what constitutes 'appropriate protection' (Oviedo *et al.*, 2005, p. 5) for intangible heritage and sacred ecologies (see Gomez *et al.*, 2010 for a good review of natural sacred sites in the context of international legal frameworks; see also Shackley, 2001; Rössler, 2004; Verschuuren *et al.*, 2010b; Goral, 2011). Such organizations include the International Union for Conservation of Nature's Specialist Group on Cultural and Spiritual Values of Protected Areas and the United Nations Educational, Scientific, and Cultural Organization (UNESCO), which has a particular focus on World Heritage Sites that have religious and/or spiritual significance.

Much of the discussion regarding natural sacred sites at the international level has been couched within the discourse of biodiversity preservation (Verschuuren *et al.*, 2010a). As noted above, natural sacred sites help to protect the world's biodiversity, and as such are important to maintain and manage from a holistic perspective. At the same time, the incorporation of religious and spiritual values in conservation approaches is also seen as a necessary component to minimize conflicts and to develop new, inclusive, and effective biocultural values (Verschuuren *et al.*, 2010a). Therefore, the creation of effective management guidelines has been the focus of several scholars and practitioners. For example, from a broad perspective, Wild and McLeod (2008) suggest that the management of natural sacred sites should include the implementation of the following six guiding principles:

- recognizing sacred natural sites already located in protected areas;
- integrating sacred natural sites within larger planning processes and management programmes;
- promoting stakeholder consent, participation, inclusion, and collaboration;
- encouraging improved knowledge and understanding of sacred sites;
- protecting sacred natural sites while providing appropriate management access and use;
- respecting the rights of sacred natural site custodians within an appropriate framework of natural policy.

Verschuuren *et al.* (2010b) also suggest several more specific actions that need to occur for better conservation of natural sacred sites:

- support the autonomous protection and management of sacred natural sites by their custodians;
- reduce/halt the threats and loss of sacred natural sites;
- support cultural revitalization and strengthen communities and their connections with their sacred natural sites;
- increase understanding and awareness of the importance and role of sacred natural sites and promote the formation of appropriate national policies and laws;
- build a body of knowledge of sacred natural sites using different ways of knowing, such as traditional indigenous knowledge, holistic science, the arts, and media;
- access and generate funding for sacred natural sites.

Verschuuren *et al.* (2017, p. 33–34) give examples of the successful balance between these international suggested guidelines and local-level issues and concerns regarding cultural, spiritual, and sacred values, including the Yolngu people in north-east Arnhem Land, Australia, who used international guidelines as a part of their Cultural Heritage Management Plan, and the Venda people in South Africa, who used these guidelines to convince the South African government to take their sacred land claims seriously.

Conclusion

Even with attempts to manage and minimize the negative impacts of pilgrimage and religious tourism at natural sacred sites and to use these sites to maintain biodiversity, there are still several concerns regarding the long-term viability and sustainability of these sites. Broader pressures beyond tourism are being placed on these sites, including population growth and pressure, urban sprawl, modernization, extractive industries and the demand for bioresources, and contestation over ownership and management, among others (Bernbaum, 2010). As such, there is growing research in this area, much of it driven by the International Union for Conservation of Nature World Commission (http://www.csvpa.org/) which has published several volumes on the protection of sacred natural sites. With increasing concerns regarding global environmental issues, such as climate change, as well as projected increases in international and domestic tourism, greater emphasis needs to be placed on how to better manage the ecological and social resilience of natural sacred sites.

What role can pilgrimage and religious tourism play in the preservation and conservation of natural sacred sites? As noted above, different faith traditions already have a strong ethic of stewardship and oneness with the natural environment. However, as noted above and in Chapters 1, 3, and 10 of this volume, there is a disconnect between religious views of the environment and the actions of pilgrims and religious tourists, who, because of their views of the sacred nature of their travels, blame other tourists and travellers for the environmental ills at sacred sites. While pilgrims and religious tourists can seek to travel in more environmentally friendly ways, this will only occur when led by faith leaders who emphasize the environmental impacts of travel to pilgrims and religious tourists and teach them how to be more environmentally sensitive during their journeys. This is the focus of the Green Pilgrimage Network (GPN) which was organized by the Alliance of Religions and Conservation in 2011 to help promote broader sustainable pilgrimage practices among different faith traditions (Singh, 2013; McIntosh, forthcoming). These practices, which align themselves well with 'slow tourism', include buying locally produced food and crafts, choosing low carbon transportation, and contributing to the greening of built sacred sites (European Green Pilgrimage Network, n.d.).

Ivakhiv (2013, pp. 85–86) argues that green pilgrimages are 'pilgrimage[s] that [are] ecologically sustainable' and 'ecological in [their] themes or orientation'. With an estimated 300–600 million people a year travelling the globe for religious purposes (Timothy, 2011; World Tourism Organization, 2011, p. xiii), any greening of pilgrimages and religious tourism would be beneficial. As natural sacred sites face continued pressures from tourism and pilgrimage development, religious leaders and organizations will hopefully continue to push the greening of pilgrimages and religious tourism and pilgrim and religious tourist behaviour that is more ecologically sustainable (Ivakhiv, 2013) in order to preserve these sites for future generations of believers and seekers.

Note

[1] Shinde (2012b) argues against this idea of pilgrimage and religious tourism being equated with 'slow tourism', whereas religious tourists tend to put as many sacred and non-sacred sites as possible in their 'sacred sightseeing' itineraries. As such, Shinde might suggest that the idea of 'slow tourism' in the context of religious tourism might relate more to those that might be labelled 'spiritual seekers' than pilgrims and religious tourists.

References

Abbott, A. (2019) Biodiversity thrives in Ethiopia's church forests. *Nature* 565(7741), 548–549.

Adler, J. (1989) Travel as performed art. *American Journal of Sociology* 94(6), 1366–1391. DOI: 10.1086/229158.

Agoramoorthy, G. (2015) Sacred rivers: their spiritual significance in Hindu religion. *Journal of Religion and Health* 54(3), 1080–1090. DOI: 10.1007/s10943-014-9934-z.

Aitpaeva, G. (2013) Introduction. In: Aitpaeva, G. (ed.) *Sacred Sites of the Southern Kyrgyzstan: Nature, Manas, Islam*. Aigine Cultural Research Center, Bishkek, Kyrgyzstan, pp. 6–10.

Albin, N. and Berwick, C. (2004) Forests and religion in Japan—from a distinctive vision of trees to a particular type of forest management. *Revue Forestière Française* 56(6), 563–572.

Alley, K.D. (2002) *On the Banks of the Ganga: When Waste Water Meets a Sacred River*. University of Michigan Press, Ann Arbor, Michigan.

Altman, N. (2002) *Sacred Water: The Spiritual Source of Life*. HiddenSpring, New Jersey.

Anderson, D.M., Salick, J., Moseley, R.K. and Xiaokun, O. (2005) Conserving the sacred medicine mountains: a vegetation analysis of Tibetan sacred sites in Northwest Yunnan. *Biodiversity and Conservation* 14(13), 3065–3091. DOI: 10.1007/s10531-004-0316-9.

Andriamarovololona, M.M. and Jones, L.P.G. (2012) The role of taboos and traditional beliefs in aquatic conservation in Madagascar. In: Pungetti, G., Oviedo, G. and Hooke, D. (eds) *Sacred Species and Sites: Advances in Biocultural Conservation*. Cambridge University Press, Cambridge, pp. 207–218.

Archer, B., Cooper, C. and Ruhanen, L. (2005) The positive and negative impacts of tourism. In: Theobald, W.F. (ed.) *Global Tourism*, 3rd edn. Elsevier, Burlington, Massachusetts, pp. 79–102.

Bacci, M., Bianchi, G., Campana, S. and Fichera, G. (2012) Historical and archaeological analysis of the Church of the Nativity. *Journal of Cultural Heritage* 13(4), e5–e26. DOI: 10.1016/j.culher.2012.10.003.

Barrow, E.G.C. (2010) Falling between the 'cracks' of conservation and religion: The role of stewardship for sacred trees and groves. In: Verschuuren, B., Wild, R., McNeely, J.A. and Oviedo, G. (eds) *Sacred Natural Sites: Conserving Nature & Culture*. Earthscan, London, pp. 42–52.

Basset, V. (2016) New Age tourism in Wirikuta: conflicts and rituals. In: Labate, B.C. and Cavnar, C. (eds) *Peyote: History, Tradition, Politics, and Conservation*. Praeger, Santa Barbara, California, pp. 191–209.

Battiste, M. and Henderson, J.Y. (2000) *Protecting Indigenous Knowledge and Heritage: A Global Challenge*. Purich Publishers, Saskatoon.

Bernbaum, E. (2006) Sacred Mountains: themes and Teachings. *Mountain Research and Development* 26(4), 304–309. DOI: 10.1659/0276-4741(2006)26[304:SMTAT]2.0.CO;2.

Bernbaum, E. (2010) Sacred Mountains and global changes: impacts and responses. In: Verschuuren, B., Wild, R., McNeely, J.A. and Oviedo, G. (eds) *Sacred Natural Sites: Conserving Nature & Culture*. Earthscan, London, pp. 33–41.

Bernbaum, E. (2012) Sacred mountains and national parks: spiritual and cultural values as a foundation for environmental conservation. In: Pungetti, G., Oviedo, G. and Hooke, D. (eds) *Sacred Species and Sites: Advances in Biocultural Conservation*. University Press, Cambridge, Cambridge, UK, pp. 83–96.

Berry, E. (2013) Religious environmentalism and environmental religion in America. *Religion Compass* 7(10), 454–466. DOI: 10.1111/rec3.12065.

Bhagwat, S.A. (2009) Ecosystem services and sacred natural sites: reconciling material and non-material values in nature conservation. *Environmental Values* 18(4), 417–427. DOI: 10.3197/096327109X125 32653285731.

Bhagwat, S.A. and Rutte, C. (2006) Sacred groves: potential for biodiversity management. *Frontiers in Ecology and the Environment* 4(10), 519–524. DOI: 10.1890/1540-9295(2006)4[519:SGPFBM]2.0.CO;2.

Bianchi, R.V. (2009) The 'Critical Turn' in Tourism Studies: A Radical Critique. *Tourism Geographies* 11(4), 484–504.

Biel, A. and Nilsson, A. (2005) Religious values and environmental concern: harmony and detachment. *Social Science Quarterly* 86(1), 178–191. DOI: 10.1111/j.0038-4941.2005.00297.x.

Bistcotti, D. and Biggart, N.W. (2014) Interfaith social change organizations in the religious-environmental movement. *Religion and Organization Theory* 41, 413–439.

Blake, K.S. (2001) Contested landscapes of Navajo sacred mountains. *The North American Geographer* 3(1), 29–62.

Blamey, R.K. (1997) Ecotourism: the search for an operational definition. *Journal of Sustainable Tourism* 5(2), 109–130. DOI: 10.1080/09669589708667280.

Blatt, W. (2005) Holy River and Magic Mountain: Public lands management and the rediscovery of the "sacred in nature". *Law & Society Review* 39(3), 681–688. DOI: 10.1111/j.1540-5893.2005.00238.x.

Bremer, T.S. (forthcoming) The religious and spiritual appeal of national parks. In: Timothy, D.J. and Olsen, D.H. (eds) *The Routledge Handbook of Religious and Spiritual Tourism*. Routledge, London and New York.

Carpenter, K.A. (2006) The interests of peoples in the cooperative management of sacred sites. *Tulsa Law Review* 42(1), 37–56.

Cennamo, P., Montuori, N., Trojsi, G., Fatigati, G. and Moretti, A. (2016) Biofilms in churches built in grottoes. *Science of The Total Environment* 543, 727–738. DOI: 10.1016/j.scitotenv.2015.11.048.

Chadha, H. and Onkar, P. (2016) Changing cities in the perspective of religious tourism – a case of Allahabad. *Procedia Technology* 24, 1706–1713. DOI: 10.1016/j.protcy.2016.05.200.

Chan, K.M.A., Balvanera, P., Benessaiah, K., Chapman, M., Díaz, S. *et al.* (2016) Opinion: why protect nature? Rethinking values and the environment. *Proceedings of the National Academy of Sciences* 113(6), 1462–1465. DOI: 10.1073/pnas.1525002113.

Chan, C., Khan, F.N. and Awan, S. (2019) Bigger issues in a smaller world: the future of the commons. In: Hudson, B., Rosenbloom, J. and Cole, D. (eds) *Routledge Handbook of the Study of the Commons*. Routledge, London and New York, pp. 401–411.

Choe, J. and Hitchcock, M. (2018) Pilgrimage to mount bromo, Indonesia. In: Olsen, D.H. and Trono, A. (eds) *Religious Pilgrimage Routes and Trails*. CAB International, Wallingford, UK, pp. 180–195.

Coleman, D. (2016) Indigenous place and diaspora space: of literalism and abstraction. *Settler Colonial Studies* 6(1), 61–76. DOI: 10.1080/2201473X.2014.1000913.

Colopy, C. (2012) *Dirty, Sacred Rivers: Confronting South Asia's Water Crisis*. Oxford University Press, New York.

Croall, J. (1995) *Preserve or Destroy: Tourism and the Environment*. Calouste Gulbenkian Foundation, London.

Curry-Roper, J.M. (1990) Contemporary Christian eschatologies and their relation to environmental stewardship. *The Professional Geographer* 42(2), 157–169. DOI: 10.1111/j.0033-0124.1990.00157.x.

Das, P. and Tamminga, K.R. (2012) The Ganges and the GAP: an assessment of efforts to clean a sacred river. *Sustainability* 4(8), 1647–1668. DOI: 10.3390/su4081647.

De Silva, P. (1998) *Environmental Philosophy and Ethics in Buddhism*. Macmillan Press, Basingstoke.

Díaz, J.L. (1977) Ethnopharmacology of sacred psychoactive plants used by the Indians of Mexico. *Annual Review of Pharmacology and Toxicology* 17(1), 647–675. DOI: 10.1146/annurev.pa.17.040177.003243.

Douros, W.J. (2012) Sacred species of national marine sanctuaries of the United States' West Coast. In: Pungetti, G., Oviedo, G. and Hooke, D. (eds) *Sacred Species and Sites: Advances in Biocultural Conservation*. Cambridge University Press, Cambridge, UK, pp. 219–230.

Dudley, N., Bhagwat, S., Higgins-Zogib, L., Lassen, B., Verschurren, B. *et al.* (2010) Conservation of biodiversity in sacred natural sites in Asia and Africa: a review of the scientific literature. In: Verschuuren, B., Wild, R., Oviedo, J.A. and McNeely, G. (eds) *Sacred Natural Sites: Conserving Nature & Culture*. Earthscan, London, pp. 19–32.

Dunlap, T.R. (2006) Environmentalism, a secular faith. *Environmental Values* 15(3), 321–330. DOI: 10.3197/096327106778226284.

Ehrenfeld, D. and Bentley, P.J. (1985) Judaism and the practice of stewardship. *Judaism* 34(3), 301–311.

El Hanandeh, A. (2013) Quantifying the carbon footprint of religious tourism: the case of hajj. *Journal of Cleaner Production* 52, 53–60. DOI: 10.1016/j.jclepro.2013.03.009.

Ellingson, S., Woodley, V.A. and Paik, A. (2012) The structure of religious environmentalism: movement organizations, interorganizational networks, and collective action. *Journal for the Scientific Study of Religion* 51(2), 266–285. DOI: 10.1111/j.1468-5906.2012.01639.x.

European Green Pilgrimage Network (n.d.) Welcome. Available at: https://www.greenpilgrimageeurope.net/ (accessed 18 December 2019).

Ferber, M. (2010) *Romanticism: A Very Short Introduction*. Oxford University Press, Oxford and New York.

Fernández-Llamazares, Á., López-Baucells, A., Rocha, R., Andriamitandrina, S.F.M., Andriatafika, Z.E. *et al.* (2018) Are sacred caves still safe havens for the endemic bats of Madagascar? *Oryx* 52(2), 271–275. DOI: 10.1017/S0030605317001648.

Fleming, W.R. and Toepper, L. (1990) Economic impact studies: relating the positive and negative impacts to tourism development. *Journal of Travel Research* 29(1), 35–42. DOI: 10.1177/004728759002900108.

Flood, G. (1966) *An Introduction to Hinduism*. Cambridge University Press, Cambridge.

Foley, R. (2011) Performing health in place: the holy well as a therapeutic assemblage. *Health & Place* 17(2), 470–479. DOI: 10.1016/j.healthplace.2010.11.014.

Frascaroli, F. (2013) Catholicism and conservation: the potential of sacred natural sites for biodiversity management in central Italy. *Human Ecology* 41(4), 587–601. DOI: 10.1007/s10745-013-9598-4.

Gábrišová, R. (2014) Ecosophy and the New Age movement in the transmodern society. *Studia Philosophiae Christianae* 50(3), 145–157.

Gadgil, M. (2018) Sacred groves: an ancient tradition of nature conservation. Scientific American. Available at: https://www.scientificamerican.com/article/sacred-groves-an-ancient-tradition-of-nature-conservation/ (accessed 26 November 2019).

pp Gardner, G. (2003) Engaging religion in the quest for a sustainable world. In: Gardner, G., Bright, C. and Starke, L. (eds) *State of the World 2003: A Worldwatch Institute Report on Progress Toward a Sustainable Society*. W.W. Norton & Company, London.

Gauthier, F., Martikainen, T. and Woodhead, L. (2013) Acknowledging a global shift: a primer for thinking about religion in consumer societies. *Implicit Religion* 16(3), 261–276. DOI: 10.1558/imre.v16i3.261.

Gesler, W.M. and Pierce, M. (2000) Hindu Varanasi. *Geographical Review* 90(2), 222–237. DOI: 10.1111/j.1931-0846.2000.tb00332.x.

Gomez, F., Heimstra, W. and Verschuuren, B. (2010) A law on sacred sites in Guatemala. *Policy Matters* 17, 119–123.

Goral, A. (2011) Research on cultural tourism development in sacral and spiritual sites from the UNESCO World Heritage List. *International Journal of Heritage and Sustainable Development* 1, 49–59.

Gottlieb, R.S. (2010) Religion and the environment. In: Gottlieb, R.S. (ed.) *Religion and the Environment*. Vol. 1. Routledge, New York.

Griffin-Pierce, T. (1997) 'When I am lonely the mountains call me': The impact of sacred geography on Navajo psychological wellbeing. *American Indian and Alaska Native Mental Health Research* 7(3), 1–10. DOI: 10.5820/aian.0703.1997.1.

Guth, J.L., Green, J.C., Kellstedt, L.A. and Smidt, C.E. (1995) Faith and the environment: religious beliefs and attitudes on environmental policy. *American Journal of Political Science* 39(2), 364–382. DOI: 10.2307/2111617.

Hall, M. (2013) The natural science ontology of environment. In: Fennell, D. (ed.) *The Routledge Handbook of Tourism and the Environment*. Routledge, London and New York, pp. 6–18.

Harper, C.L. (2008) Religion and the environment. *Journal of Religion & Society, Supplemental Series* 3, 5–26.

Heintzman, P. (2009) Nature-based recreation and spirituality: a complex relationship. *Leisure Sciences* 32(1), 72–89. DOI: 10.1080/01490400903430897.

Helms, M.W. (2002) Sacred landscape in the early medieval cloister: utility, paradise, and the cosmic mountain. *Anthropos* 97(2), 435–453.

Holden, A. (2003) In need of new environmental ethics for tourism? *Annals of Tourism Research* 30(1), 94–108. DOI: 10.1016/S0160-7383(02)00030-0.

Holden, A. (2009) The environment-tourism nexus: influence of market ethics. *Annals of Tourism Research* 36(3), 373–389.

Hooker, A.M. (1994) American Indian sacred sites on federal public lands: resolving conflicts between religious use and multiple use at El Malpais national monument. *American Indian Law Review* 19(1), 133–158.

Howard, C. (2012) Speeding up and slowing down: pilgrimage and slow travel through time. In: Fullagar, S., Markwell, K. and Wilson, E. (eds) *Slow Tourism: Experiences and Mobilities*. Channel View Publications, Bristol, UK, pp. 11–24.

Hughes, J.D. and Chandran, M.D.S. (1998) Sacred groves around the earth: an overview. In: Ramakrishnan, P.S., Saxena, K.G. and Chandrashekara, U.M. (eds) *Conserving the Sacred for Biodiversity Management*. Oxford and India Book House, New Delhi, India.

Huntsinger, L. and Fernández-Giménez, M. (2000) Spiritual pilgrims at Mount Shasta, California. *Geographical Review* 90(4), 536–558. DOI: 10.1111/j.1931-0846.2000.tb00353.x.

Ivakhiv, A. (2003) Nature and self in New Age pilgrimage. *Culture and Religion* 4(1), 93–118. DOI: 10.1080/01438300302812.

Ivakhiv, A. (2013) Green pilgrimage: problems and prospects for ecology and peace-building. In: Pazos, A.M. (ed.) *Pilgrims and Pilgrimages as Peacemakers in Christianity, Judaism and Islam*. Ashgate, Surrey, pp. 85–103.

Jenkins, W. (2013) *Ecologies of Grace: Ethnics and Christian Theology*. Oxford University Press, Oxford and New York.

John, H. (2019) Sacred groves of North Kerala: the last refuge for biodiversity amongst urbanisation. Mongabay. Available at: https://india.mongabay.com/2019/01/sacred-groves-of-north-kerala-the-last-refuge-for-biodiversity-amongst-urbanisation/ (accessed 29 November 2019).

Jones, P. and Pennick, N. (1995) *A History of Pagan Europe*. Routledge, London and New York.

Kaehlig, C.B., Wright, A. and Smith, C. (2012) *Volcanoes of Indonesia: Creators and Destroyers*. Editions Didier Millet, Singapore.

Kato, K. and Progano, R.N. (2017) Spiritual (walking) tourism as a foundation for sustainable destination development: Kumano-kodo pilgrimage, Wakayama, Japan. *Tourism Management Perspectives* 24, 243–251. DOI: 10.1016/j.tmp.2017.07.017.

Khan, M.L., Khumbongmayum, A.D. and Tripathi, R.S. (2008) The sacred groves and their significance in conserving biodiversity: an overview. *International Journal of Ecology and Environmental Sciences* 34(3), 277–291.

Kiernan, K. (2015) Landforms as sacred places: implications for geodiversity and geoheritage. *Geoheritage* 7(2), 177–193. DOI: 10.1007/s12371-014-0128-6.

King, E.I.O., Viji, C. and Narasimhan, D. (1997) Sacred groves: traditional ecological heritage. *International Journal of Ecology and Environmental Sciences* 23(4), 463–470.

Kuokkanen, R. (2006) Indigenous peoples on two continents. *Native American Studies* 20(2), 1–6.

Lane, M.B., Ross, H., Dale, A.P. and Rickson, R.E. (2003) Sacred land, mineral wealth, and biodiversity at Coronation Hill, Northern Australia: Indigenous knowledge and SIA. *Impact Assessment and Project Appraisal* 21(2), 89–98. DOI: 10.3152/147154603781766374.

Leeming, D.A. (2010) *Creation Myths of the World: An Encyclopedia*, 2nd edn. ABC-CLIO, Santa Barbara, California.

Leiper, N. (1979) The framework of tourism: towards a definition of tourism, tourists, and the tourist industry. *Annals of Tourism Research* 6(4), 390–407.

Liljeblad, J. and Verschuuren, B. (2019) *Indigenous Perspectives on Sacred Natural Sites: Culture, Governance and Conservation*. Routledge, London and New York.

Logan, P. (1980) *The Holy Wells of Ireland*. Smythe, Buckinghamshire, UK.

Lourdunathan, S. (2012) Ecosophical concerns in the Sikh tradition. In: Cain, C.C. (ed.) *In Many Heavens, One Earth: Readings on Religion and the Environment*. Lexington Books, Lanham, pp. 165–169.

Lowenstein, T. (1993) *Ancient Land: Sacred Whale. The Inuit Hunt and its Rituals*. Bloomsbury, London.

Lusby, C. (2017) Hard and soft tourism. In: Lusby, C. (ed.) *The International Encyclopedia of Travel and Tourism*. SAGE, Thousand Oaks, California, pp. 566–568.

Lüthi, D. (2010) Private cleanliness, public mess: purity, pollution and space in Kottar, South India. In: Dürr, E. and Jaffe, R. (eds) *Urban Pollution: Cultural Meanings, Social Practices*. Berghahn, New York, pp. 57–85.

Matose, F., Oyono, P.R. and Murombedzi, J. (2019) Customary authority and commons governance. In: Hudson, B., Rosenbloom, J. and Cole, D. (eds) *Routledge Handbook of the Study of the Commons*. Routledge, London and New York, pp. 334–344.

Mazumdar, S. and Mazumdar, S. (1993) Sacred space and place attachment. *Journal of Environmental Psychology* 13(3), 231–242. DOI: 10.1016/S0272-4944(05)80175-6.

McIntosh, I. (forthcoming) Pilgrimage and peace-building. In: Olsen, D.H. and Timothy, D.J. (eds) *The Routledge Handbook of Religious and Spiritual Tourism*. Routledge, London and New York.

McIvor, A. and Pungetti, G. (2012) The conservation status of sacred species: a preliminary study. In: Pungetti, G., Oviedo, G. and Hooke, D. (eds) *Sacred Species and Sites: Advances in Biocultural Conservation*. Cambridge University Press, Cambridge, UK, pp. 195–206.

McWilliam, A. (2009) The spiritual commons: some immaterial aspects of community economies in eastern Indonesia. *The Australian Journal of Anthropology* 20(2), 163–177. DOI: 10.1111/j.1757-6547.2009.00024.x.

Mill, R.C. and Morrison, A.M. (2002) *The Tourism System*. Kendall Hunt, Dubuque, Iowa.

Muir, S. (2007) The good of New Age goods: Commodified images of Aboriginality in New Age and alternative spiritualities. *Culture and Religion* 8(3), 233–253.

Nan, W., Garrett, L. and McGowan, P. (2012) Pheasant conservation, sacred groves and local culture in Sichuan, China. In: Pungetti, G., Oviedo, G. and Hooke, D. (eds) *Sacred Species and Sites: Advances in Biocultural Conservation*. Cambridge University Press, Cambridge, pp. 265–277.

Ndoro, W. and Wijesuriya, G. (2015) Heritage management and conservation: from colonization to globalization. In: Keskell, L. (ed.) *Global Heritage: A Reader*. Wiley, Chichester, UK, pp. 131–149.

Ntiamoa-Baidu, Y. (1995) *Indigenous vs. Introduced Biodiversity Conservation Strategies: The Case of Protected Area Systems in Ghana*. WWF-Biodiversity Support Program, Washington, DC.

Nyamweru, C. and Kimaru, E. (2008) The contribution of ecotourism to the conservation of natural sacred sites: a case study from coastal Kenya. *Journal for the Study of Religion, Nature and Culture* 2(3), 327–350. DOI: 10.1558/jsrnc.v2i3.327.

Olsen, D.H. (2017) Other journeys of creation: non-representational theory, co-creation, failure, and the soul. *Tourism Recreation Research* 42(1), 120–124. DOI: 10.1080/02508281.2016.1261782.

Olsen, D.H. (2019) Religion, spirituality, and pilgrimage in a globalising world. In: Timothy, D.J. (ed.) *Handbook of Globalisation and Tourism*. Edward Elgar, London, pp. 270–283.

O'Neil, D.J. (2001) The New Age movement and its societal implications. *International Journal of Social Economics* 28(5/6/7), 456–475. DOI: 10.1108/03068290110360795.

Ormsby, A.A. (2011) The impacts of global and national policy on the management and conservation of sacred Groves of India. *Human Ecology* 39(6), 783–793. DOI: 10.1007/s10745-011-9441-8.

Oviedo, G. and Jeanrenaud, S. (2007) Protecting sacred natural sites of Indigenous and traditional peoples. In: Mallarach, J.-M. and Papayanni, T. (eds) *Protected Areas and Spirituality: Proceedings of the First Workshop of The Delos Initiative., The World Conservation Union (IUCN)*. Gland, Switzerland, pp. 77–99.

Oviedo, G., Jeanrenaud, S. and Otegui, M. (2005) Protecting sacred natural sites of Indigenous and traditional peoples: an IUCN perspective. Available at: https://www.iucn.org/sites/dev/files/import/downloads/sp_protecting_sacred_natural_sites_indigenous.pdf (accessed 27 November 2019).

Pandy, D.N. (2000) Sacred water and sanctified vegetation: tanks and trees in India. Available at: https://dlc.dlib.indiana.edu/dlc/bitstream/handle/10535/2384/pandeyd041700.pdf?sequence=1&isAllowed=y (accessed 2 December 2019).

Papayannis, T. and Mallarach, J. (2008) *Proceedings of the Second Delos Workshop*, Ouranoupoli, Greece, October 2008.

Pommaret, F. (2009) *Bhutan: Himalayan Mountain Kingdom*. Odyssey Books & Guides, Hong Kong.

Powici, C. (2004) What is wilderness? John Muir and the question of the wild. *Scottish Studies Review* 5(1), 74–86.

Pungetti, G. (2012) Sacred species and sites: Dichotomies, concepts and new directions in biocultural diversity conservation. In: Pungetti, G., Oviedo, G. and Hooke, D. (eds) *Sacred Species and Sites: Advances in Biocultural Conservation*. Cambridge University Press, Cambridge, pp. 13–27.

Pungetti, G. (2013) Biocultural diversity for sustainable ecological, cultural and sacred landscapes: the biocultural landscape approach. In: Fu, B. and Jones, B. (eds) *Landscape Ecology for Sustainable Environment and Culture*. Springer, Dordrecht, pp. 55–76.

Ramakrishnan, P.S. and Harmonand, D. (2003) Conserving the sacred: the protective impulse and the origins of modern protected areas. In: Harmonand, D. and Putney, A.D. (eds) *The Full Value of Parks*. Rowman & Littlefield, Lanham, Maryland, pp. 27–42.

Rattue, J. (1995) *The Living Stream: Holy Wells in Historical Context*. The Boydell Press, Woodbridge, UK.

Redden, G. (2005) The New Age: towards a market model. *Journal of Contemporary Religion* 20(2), 231–246. DOI: 10.1080/13537900500067851.

Reinhard, J. (1985) Sacred Mountains: an ethno-archaeological study of high Andean ruins. *Mountain Research and Development* 5(4), 299–317. DOI: 10.2307/3673292.

Roberts, M. (2012) Genealogy of the sacred: Maori beliefs concerning lizards. In: Pungetti, G., Oviedo, G. and Hooke, D. (eds) *Sacred Species and Sites: Advances in Biocultural Conservation*. Cambridge University Press, Cambridge, UK, pp. 249–264.

Robinson, G. (1930) Some cave chapels of southern Italy. *The Journal of Hellenic Studies* 50(2), 186–209. DOI: 10.2307/626810.

Rössler, M. (2004) Managing world heritage cultural landscapes and sacred sites. In: de Merode, E., Smeets, R. and Westrik, C. (eds) *World Heritage Papers*. 13. Paris, UNESCO World Heritage Center, pp. 45–48.

Rountree, K. (2006) Performing the divine: Neo-pagan pilgrimages and embodiment at sacred sites. *Body & Society* 12(4), 95–115. DOI: 10.1177/1357034X06070886.

Rutte, C. (2011) The sacred commons: conflicts and solutions of resource management in sacred natural sites. *Biological Conservation* 144(10), 2387–2394. DOI: 10.1016/j.biocon.2011.06.017.

Saarinen, J. (2006) Traditions of sustainability in tourism studies. *Annals of Tourism Research* 33(4), 1121–1140. DOI: 10.1016/j.annals.2006.06.007.

Salick, J., Amend, A., Anderson, D., Hoffmeister, K., Gunn, B. *et al.* (2007) Tibetan sacred sites conserve old growth trees and cover in the eastern Himalayas. *Biodiversity and Conservation* 16(3), 693–706. DOI: 10.1007/s10531-005-4381-5.

Samakov, A. and Berkes, F. (2017) Spiritual commons: sacred sites as core of community-conserved areas in Kyrgyzstan. *International Journal of the Commons* 11(1), 422–444. DOI: 10.18352/ijc.713.

Schlosberg, D. and Carruthers, D. (2010) Indigenous struggles, environmental justice, and community capabilities. *Global Environmental Politics* 10(4), 12–35. DOI: 10.1162/GLEP_a_00029.

Semwal, N. and Akolkar, P. (2006) Water quality assessment of sacred Himalayan rivers of Uttaranchal. *Current Science* 91(4), 486–496.

Seshardi, K.S. and Ganesh, T. (2011) Faunal mortality on roads due to religious tourism across time and space in protected areas: a case study from South India. *Forest Ecology and Management* 262(9), 1713–1721.

Sessions, G. (1987) The deep ecology movement: a review. *Environmental History Review* 11(2), 105–125.

Shackley, M. (2001) Sacred world heritage sites: balancing meaning with management. *Tourism Recreation Research* 26(1), 5–10. DOI: 10.1080/02508281.2001.11081171.

Sharma, N. (2020) Interpreting the sacred in dark tourism. In: Olsen, D.H. and Korstanje, M. (eds) *Dark Tourism and Pilgrimage*. CAB International, Wallingford, UK, pp. 25–37.

Sharpley, R. (2009) *Tourism Development and the Environment: Beyond Sustainability?* Earthscan, Abingdon, UK.

Sharpley, R. (2011) *The Study of Tourism: Past Trends and Future Directions*. Routledge, New York.

Sharpley, R. and Jepson, D. (2011) Rural tourism: a spiritual experience? *Annals of Tourism Research* 38(1), 52–71.

Shengji, P. and Wild, R. (2010) The road to the future? The biocultural values of the Holy Hill forests of Yunnan Province, China. In: Verschuuren, B., Wild, R., McNeely, J.A. and Oviedo, G. (eds) *Sacred Natural Sites: Conserving Nature & Culture*. Earthscan, London, pp. 98–106.

Shibley, M.A. and Wiggins, J.L. (1997) The greening of mainline American religion: a sociological analysis of the environmental ethics of the national religious partnership for the environment. *Social Compass* 44(3), 333–348. DOI: 10.1177/003776897044003003.

Shinde, K.A. (2007) Pilgrimage and the environment: challenges in a pilgrimage centre. *Current Issues in Tourism* 10(4), 343–365. DOI: 10.2167/cit259.0.

Shinde, K.A. (2010) Entrepreneurship and indigenous entrepreneurs in religious tourism in India. *International Journal of Tourism Research* 12(5), 523–535.

Shinde, K.A. (2012a) Policy, planning, and management for religious tourism in Indian pilgrimage sites. *Journal of Policy Research in Tourism, Leisure and Events* 4(3), 277–301. DOI: 10.1080/19407963.2012.726107.

Shinde, K.A. (2012b) From route to road and body to car: changing aesthetics of religious tourism in a sacred landscape. *Literature & Aesthetics* 22(1), 88–108.

Shinde, K.A. (2018) Palkhi: a moving sacred town. In: Olsen, D.H. and Trono, A. (eds) *Religious Pilgrimage Routes and Trails*. CAB International, Wallingford, UK, pp. 150–166.

Singh, N-G.K. (ed.) (1995) *The Name of My Beloved: Verses of the Sikh Gurus*, Trans. N-G.K. Singh. Penguin, New Delhi, India.

Singh, S. (2012) Slow travel and Indian culture: philosophical and practical aspects. In: Fullagar, S., Markwell, K. and Wilson, E. (eds) *Slow Tourism: Experiences and Mobilities*. Channel View Publications, Bristol, UK, pp. 214–226.

Singh, R.P.B. (2013) *Hindu Tradition of Pilgrimage: Sacred Space & System*. Dev Publishers & Distributors, Delhi, India.

Singh, T.V. and Kaur, J. (1986) Paradox of mountain tourism: case references from the Himalaya. *Tourism Recreation Research* 11(1), 19–28. DOI: 10.1080/02508281.1986.11014413.

Smith, A. and Pulver, S. (2009) Ethics-based environmentalism in practice: Religious-environmental organizations in the United States. *Worldviews: Global Religions, Culture, and Ecology* 13(2), 145–179. DOI: 10.1163/156853509X438580.

Sobrevila, C. (2008) *The Role of Indigenous Peoples in Biodiversity Conservation*. World Bank, Washington, DC.

Sponsel, L.E. (2015) Sacred caves of the world: illuminating the darkness. In: Brunn, S.D. (ed.) *The Changing World Religion Map*. Springer, Dordrecht, The Netherlands, pp. 503–522.

Sponsel, L.E. and Natadecha-Sponsel, P. (2004) Illuminating darkness: the monk-cave-bat-ecosystem complex in Thailand. In: Gottlieb, R.S. (ed.) *This Sacred Earth: Religion, Nature, Environment*. Routledge, New York, pp. 134–144.

Stewart-Harawira, M. (2012) Returning the sacred: Indigenous ontologies in perilous times. In: Williams, L., Roberts, R. and Mcintosh, A. (eds) *Radical Human Ecology: Intercultural and Indigenous Approaches*. Routledge, London and New York, pp. 73–85.

Stuart-Fox, D.J. (2002) *Pura Besakih: Temple, Religion and Society in Bali*. KITLV, Leiden.

Timothy, D.J. (2011) *Cultural and Heritage Tourism: An Introduction*. Channel View Publication, Bristol, UK.

Timothy, D.J. (2012) Religious views of the environment: Sanctification of nature and implications for tourism. In: Holden, A. and Fennell, D.A. (eds) *The Routledge Handbook of Tourism and the Environment*. London and New York, Routledge, pp. 53–64.

Timothy, D.J. and Boyd, S.W. (2015) *Tourism and Trails: Cultural, Ecological and Management Issues*. Channel View Publications, Bristol, UK.

Timothy, D.J. and Conover, P.J. (2006) Nature religion, self-spirituality, and new age tourism. In: Timothy, D.J. and Olsen, D.H. (eds) *Tourism, Religion, and Spiritual Journeys*. Routledge, London and New York, pp. 139–155.

Trombino, G. and Trono, A. (2018) Environment and sustainability as related to religious pilgrimage routes and trails. In: Olsen, D.H. and Trono, A. (eds) *Religious Pilgrimage Routes and Trails*. CAB International, Wallingford, UK, pp. 49–60.

Tucker, M.E. and Grim, J.A. (2001) Introduction: the emerging alliance of world religions and ecology. *Daedalus* 130(4), 1–22.

Varghese, K.J., Anila, J., Nagalekshmi, R., Resiya, S. and Sonu, J. (2010) Dasapushpam: the traditional uses and the therapeutic potential of ten sacred plants of Kerala state in India. *International Journal of Pharmaceutical Sciences and Research* 1(10), 50–59.

Varner, G.R. (2009) *Sacred Wells: A Study in the History, Meaning, and Mythology of Holy Wells & Waters*. Algora Publishing, New York.

Verschuuren, B. (2010a) Arguments for developing biocultural conservation approaches for sacred natural sites. In: Verschuuren, B., Wild, R., McNeely, J.A. and Oviedo, G. (eds) *Sacred Natural Sites: Conserving Nature & Culture*. Earthscan, London, pp. 62–71.

Verschuuren, B., Wild, R., McNeely, J.A. and Oviedo, G. (2010b) *Sacred Natural Sites: Conserving Nature & Culture*. Earthscan, London.

Verschuuren, B., Wild, R., McNeely, J.A. and Oviedo, G. (2010c) Introduction: sacred natural sites the foundations of biodiversity. In: Verschuuren, B., Wild, R., McNeely, J.A. and Oviedo, G. (eds) *Sacred Natural Sites: Conserving Nature & Culture*. Earthscan, London, pp. 1–13.

Verschuuren, B., Wild, R. and Verschoor, G. (2017) Connecting policy and practice for the conservation of sacred natural sites. In: Verschuuren, B. and Brown, S. (eds) *Cultural and Spiritual Significance of Nature in Protected Areas: Governance, Management and Policy*. Routledge, New York, pp. 11–40.

Vogel, D. (2001) How green is Judaism? Exploring Jewish environmental ethics. *Business Ethics Quarterly* 11(2), 349–363. DOI: 10.2307/3857753.

Wall, G. and Mathieson, A. (2006) *Tourism: Change, Impacts and Opportunities*. Pearson, Harlow, UK.

Wang, B. (2019) Sacred trash and personhood: living in daily waste-management infrastructures in the Eastern Himalayas. *Cross-Currents: East Asian History and Culture Review* 8(1), 224–248. DOI: 10.1353/ach.2019.0009.

Ward, W.E. (1952) The lotus symbol: its meaning in Buddhist art and philosophy. *The Journal of Aesthetics and Art Criticism* 11(2), 135–146. DOI: 10.2307/426039.

Wearing, S.L., McDonald, M. and Ankor, J. (2016) Journeys of creation: experiencing the unknown, the other and authenticity as an epiphany of the self. *Tourism Recreation Research* 41(2), 157–167. DOI: 10.1080/02508281.2016.1159058.

Wei, J. (2004) Humanistic advice of the Buddha to political leaders. *The Hsi Lai Journal of Humanistic Buddhism* 5, 481–489.

Weise, K. (2013) Managing the sacred garden of Lumbini. In: Silva, K.D. and Chapagain, N.K. (eds) *Asian Heritage Management: Contents, Concerns and Prospects*. Routledge, London, pp. 122–136.

Westervelt, W.D. (1916) *Hawaiian Legends of Volcanoes*. Ellis Press, Boston.

Wheatley, B.P. (1999) *The Sacred Monkeys of Bali*. Waveland Press, Long Grove, IL.

White, L. (1967) The historical roots of our ecologic crisis. *Science* 155(3767), 1203–1207. DOI: 10.1126/science.155.3767.1203.

Whitfield, R., Whitfield, S. and Agnew, N. (2000) *Cave Temples of Mogao: Art and History on the Silk Road*. Getty Publications, Los Angeles.

Wild, R. and McLeod, C. (2008) *Sacred Natural Sites: Guidelines for Protected Area Managers*. IUCN/UNESCO, Gland, Switzerland and Paris.

Winter, C. (2007) The intrinsic, instrumental and spiritual values of natural area visitors and the general public: a comparative study. *Journal of Sustainable Tourism* 15(6), 599–614. DOI: 10.2167/jost686.0.

Woodrum, E. and Wolkomir, M.J. (1997) Religious effects on environmentalism. *Sociological Spectrum* 17(2), 223–234. DOI: 10.1080/02732173.1997.9982161.

World Tourism Organization (2011) *Religious Tourism in Asia and the Pacific*. World Tourism Organization, Madrid, Spain.

Woosnam, K.M., Aleshinloye, K.D. and Maruyama, N. (2016) Solidarity at the Osun Osogbo sacred grove—a UNESCO world heritage site. *Tourism Planning & Development* 13(3), 274–291. DOI: 10.1080/21568316.2015.1104380.

Yerbury, R. and Boyd, W. (2018) Wild dolphins, nature and leisure: whose wellbeing? In: Carr, N. and Young, J. (eds) *Wild Animals and Leisure*. Routledge, London and New York, pp. 149–164.

York, M. (2001) New Age commodification and appropriation of spirituality. *Journal of Contemporary Religion* 16(3), 361–372. DOI: 10.1080/13537900120077177.

Zurich, D. (2002) Water temples of Bali. *Focus on Geography* 47(2), 1–8. DOI: 10.1111/j.1949-8535.2002.tb00034.x.

3 Managing the Environment in Religious Tourism Destinations: A Conceptual Model

Kiran A. Shinde*
College of Arts, Social Sciences and Commerce, La Trobe University, Melbourne, Australia

Introduction

Like other tourism niche markets, religious tourism is a growth market, with marketers using religious culture as a part of economic strategies in countries around the world (Olsen, 2003). Religious destinations are sacred places that are integral to the development and maintenance of religious and social identities (Collins-Kreiner, 2010). They are important centres for religious practices, such as pilgrimage, and for spiritual development, and can be found throughout the world, ranging from natural formations to pilgrim-towns developed around places of worship commemorating deities and gods and their miraculous powers (Bhardwaj, 1997).

While religious sites have seen an increase in pilgrim, religious tourist, and general tourist visitation in the past few decades, this increased visitation has led to change in the physical environment at these destinations (e.g. Freitag, 1989; Ambrósio, 2003; Shackley, 2004; Singh, 2004; Timothy and Olsen, 2006; Shinde, 2007, 2012a; Henderson, 2011; Stausberg, 2012). A few studies have highlighted how the production of the sacred itself drives environmental change, where natural landscapes of mythological origins are transformed into pilgrim towns and developed around temples and religious infrastructure for pilgrims and visitors (Freitag, 1989; Joseph, 1994; Lochtefeld, 2010; Shinde, 2012a). Other research has focused on the agency and authority with which different institutions and stakeholders participate in and address environmental concerns while demonstrating how environmental responsibility in places where sacred and religious values dominate as it relates to place attachment and the development of a 'sense of place' (e.g. Alley, 2002; Haberman, 2006; Shinde, 2011). These studies, however, have not led to the development of a conceptual framework that examines the interrelationships and dynamics between pilgrimage, religious tourism, and environmental impacts and their management.

Therefore, the purpose of this chapter is to do a comparative analysis of the environmental issues in six religious tourism destinations and then use this analysis as a basis to propose a conceptual model for explaining environmental management in religious tourism destinations. All the destinations selected for this analysis are from the state of Maharashtra in western India. This was done so that at least one governing parameter – legislative and institutional frameworks – was common to all sites. This commonality allowed for exploring the influence of other factors, such as the size of the town, the nature

*k.shinde@latrobe.edu.au

of the pilgrimage economy, the scale of pilgrim- or catchment-field, landscape attributes, and land uses, among others, in articulating and responding to environmental problems. Based on a review of existing approaches to management of pilgrimage sites (Shackley, 2001; Olsen, 2006; Raj and Morpeth, 2007; Henderson, 2011; Shinde, 2012b) and the findings from the above mentioned six destinations, the author suggests a conceptual model that involves the analytical categories of *environmental processes*, *institutional responsibility*, and *place attachment*, and illustrates how this model allows for a comprehensive explanation of the complexities of managing the environment in a wide variety of religious tourism destinations.

The chapter is divided into seven sections. The second discusses the existing approaches to the study of Hindu pilgrimage sites, while in the third, the criteria used in the selection of the six destinations is outlined with an emphasis on how pilgrimage and religious tourism works within these destinations. Using indicators commonly used in environmental assessments, the fourth section provides a comparative discussion of the environmental scenarios in these places. The different institutions and their responses to the environmental challenges are explained in the fifth section, while in the sixth the conceptual model is proposed and expanded upon. The final and concluding section outlines further research directions regarding the practical implementation of this model to understand and mitigate the environmental impacts of religious tourism in other destinations.

Hindu Pilgrimage Sites: A Quick Primer

Hindu pilgrimage is an ancient phenomenon with hundreds of pilgrimage trails and centres throughout India. Because of the vast number of Hindu pilgrimages and the complexity of understanding them as a system, (Vidyarthi, 1961; Vidyarthi *et al.*, 1979; see Dubey, 1995; Shinde, 2011) the *sacred complex model* has been developed, which comprises three interrelated analytical categories: sacred geography, sacred specialists, and sacred performances.

Every pilgrimage place has a *sacred geography* which includes a network of sacred spots where the natural features such as rivers, trees, and ponds are attributed with legends of the presiding deities of the place. Shrines, sanctified artefacts, and objects commemorating the presence of the divine also form the core of this sacred geography which is often demarcated by a sacred boundary. This territory becomes the main arena for performance of pilgrimage rituals (Vidyarthi *et al.*, 1979; Shinde, 2008; Lochtefeld, 2010).

Sacred specialists are persons who act as intermediaries between pilgrims and sacred objects. Several types of specialists, such as priests, religious gurus, monks, and ascetics direct or guide pilgrims in the performance of a variety of religious activities and rituals. Sacred specialists are considered to belong to the 'ritual complex' and/or the 'spiritual complex' of a place based on their functional roles. Pilgrimage priests may retain the exclusive authority to perform the rituals, while religious gurus, monks, and ascetics may impart spiritual knowledge or initiate followers into their religious sects (Caplan, 1982; Joseph, 1994; Lochtefeld, 2010).

Sacred performances include all rituals that are believed to assist pilgrims in transcending from the material world to the divine world. These performances are deeply embedded in religious beliefs and are also used for asserting social status or reinforcing social hierarchies (van Der Veer, 1988; Joseph, 1994; Shinde, 2008). For example, performing rituals for deceased ancestors on the banks of the River Ganges in Banaras is one of the most pious acts for Hindus and is the driving force in the pilgrimage economy of Banaras (Freitag, 1989). Performing rituals on auspicious occasions in the Hindu lunar calendar is particularly considered meritorious: these occasions arise out of certain cosmic alignments of planets, the sun, and the moon. Worshippers believe that at these times, when cosmic forces will be in their favour, performing rituals in service of their deities will help them acquire merit that will not only improve their life now but will also lead to a better afterlife.

Most Hindu pilgrimage sites can be explained using the sacred complex model as it encapsulates the basic need for pilgrimage and its manifestation through religious and cultural activities surrounding a deity. In many studies,

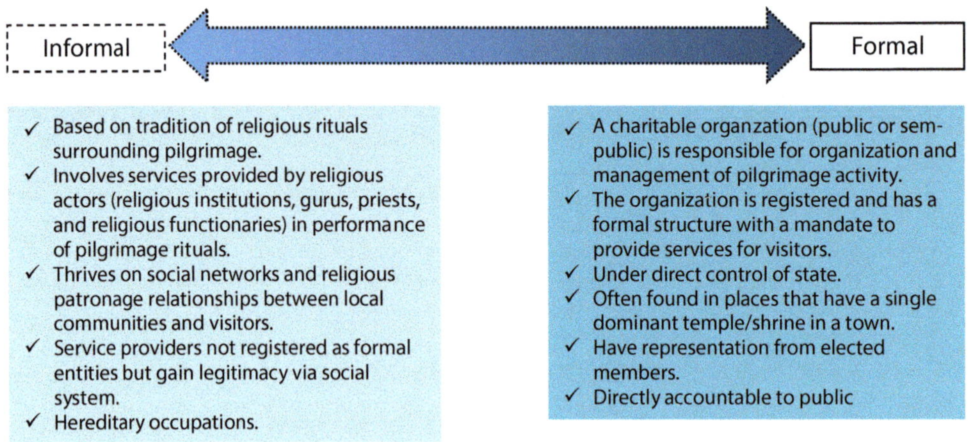

Fig. 3.1. The informal-formal axis of pilgrimage-management in Hindu pilgrimage sites.

the concept of the 'sacred' is interchanged with the concept of the 'religious', as they tend to signify similar meanings in the Hindu pantheon. This model, when used in tandem with the geographic distribution model proposed by Bhardwaj (1973), which focuses on the development of a hierarchy of catchment areas among Hindu pilgrimage sites and categorises them as pan-Indian, national, supra-regional, regional, and local, explains a wide universe of thousands of Hindu pilgrimage sites throughout India.

Hindu pilgrimage sites have complex systems of management due to the presence of both religious institutions and specialists operating in the pilgrimage economy and non-religious institutions that govern and administer the broader political context (Freitag, 1989; Lochtefeld, 2010). From a tourism perspective, one way to understand pilgrimage sites and their economies is through the interplay of what can be referred to as the 'formal-informal axis'or 'formal-informal typology' (Gladstone, 2005; Shinde, 2012b) as illustrated in Fig. 3.1.

The informal end of this axis is where the tradition of pilgrimage relies on informal networks built around social patronage relationships between the visitors and religious actors, including religious institutions, gurus, priests, and religious functionaries who help to enhance the experiential nature of pilgrimage (van Der Veer, 1988; Lochtefeld, 2010). At the formal end, the temple or shrine is administered and managed by a public organization (often under the direct control of state government) which has a mandate for providing a range of services for visitors. Hindu places of worship that have a single dominant shrine as the focus of pilgrimage are often governed through such a system (Verdia, 1982; Patnaik, 2006; Shinde, 2007). Between the two ends of the axis are several possibilities through which various actors structure their involvement in the pilgrimage economy.

The Six Destination Case Studies

The six destinations used as case studies to develop the aforementioned conceptual framework for the environmental analysis of pilgrimage sites are located in the state of Maharashtra in western India. Being in the same state provides an opportunity to do a cross-sectional analysis based on commonalities in their administration and governance, which is a crucial component for environmental management. Moreover, these destinations belong to an ethno-religious landscape shared by a large section of the population in the state (Feldhaus, 1986; Dhere, 2011). The destinations also represent a diversity of environmental contexts: two are perched atop mountains, a river runs through two sites, and two are near heavily urbanized centres. In terms of religious practice, two sites are dedicated to deities that are important for life cycle rituals; two sites are centres of devotional worship to

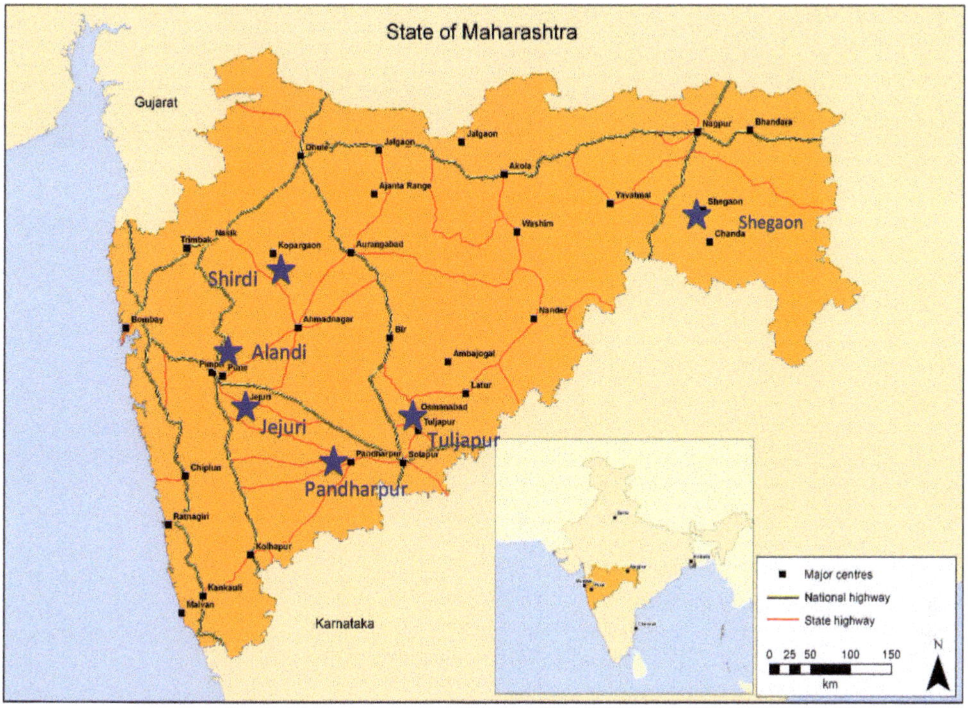

Fig. 3.2. A map showing the location of the six destinations. Drawn by the author, adapted from DIVA-GIS. http://www.diva-gis.org/gData (public domain data).

modern gurus; one is famous as the birthplace of a 13th-century poet-saint; and one is the centre of devotional worship to the regional incarnation of Vishnu (one of the trinity gods in Hindu pantheon). The sites also differ in terms of population and size and thus in urban hierarchy. Between them, the destinations represent the full spectrum of formal to informal pilgrimage economy discussed earlier. This chapter focuses on how the variable characteristics of these case studies play out in the environmental outcomes at these sites related to pilgrimage and religious tourism to see if patterns of environmental management and governance can be identified within the myriad of environmental problems within these destinations.

The six destinations of Alandi, Jejuri, Pandharpur, Shegaon, Shirdi, and Tuljapur (Fig. 3.2) were studied under a five-year research project between 2011 and 2016. Fieldwork conducted in these destinations involved interviews with key groups of stakeholders, participant observation, and an analysis of local writings on the topic. The destinations are introduced below with a focus only on contemporary aspects of pilgrimage and religious tourism as deemed relevant for environmental analysis (for a comprehensive discussion of the mythology, history, and cultural dimensions of these case study sites, see Pohnerkar, 1961; Mokashi, 1987; Rigopoulos, 1993; SGMS, 2007; Dhere, 2008).

Alandi

Alandi is a pilgrim-town dedicated to Dnyaneshwar – a 13th-century poet-saint who propagated a devotional form of worship known as *bhakti* where a devotee seeks to attain salvation by singing devotional songs in praise of the supreme deity (often Vishnu). Dnyaneshwar translated the Hindu religious treatise of Bhagavad Gita from Sanskrit into the Marathi, which language is spoken in the Maharashtra region. There are several miracle stories associated with the saint and his supreme devotion to Vitthal (or Vithoba) of Pandharpur. In Alandi, Dnyaneshwar attained *Samadhi* (i.e. the conscious departure from the physical body for union with the divine) under a tree near the Siddheshwar temple on the banks of Indrayani river. In Alandi, the most significant tradition is the *wari* – a walking procession or pilgrimage undertaken by followers of Dnyaneshwar, in which they re-enact his pilgrimage to Pandharpur by carrying his symbolic sandals in a *palkhi* (palanquin) (Mokashi, 1987). The followers, known as *warkari*, participate in this devotional tradition at least four times every year (mainly starting on Ekadashi – the eleventh lunar day of the Hindu calendar – which is dedicated to the worship of Lord Vishnu). The largest gathering of pilgrims takes place on the Ekadashi in the Hindu month of Ashada (June or July), which marks the beginning of *Chaturmas*—a period of four months when Vishnu is believed to fall asleep.

In Alandi, people follow Dnyaneshwar's format of devotional worship that includes chanting and communal singing to relate to Vitthal. As such, the ritualistic form of pilgrimage involving religious specialists, including priests and guides offering ritual services to pilgrims, has not developed in Alandi. The main temple complex was built in the 17th century, and is presently managed by the Dnyaneshwar Mandir Sansthan Trust, which was established in 1934. There are other religious sites in Alandi that pilgrims visit, including temples dedicated to other deities and the family members of Dnayeshwar. In addition, many pilgrims dip themselves in the holy river as a part of bathing rituals. Over time, many caste groups have built a basic pilgrimage infrastructure, such as pilgrim-lodges and *dharamshalas* (i.e. rest houses for spiritual pilgrims) to meet the social and cultural needs of their fellow pilgrims and to shelter widows, the destitute, mendicants, and others of low socioeconomic standing. There are presently close to 400 Dharamshalasin Alandi (Wilbur Smith Associates, 2005).

Alandi has also acquired a reputation as a sacred place where people go to get married in the divine presence of the saint by following a very simple exchange of vows and rituals. This becomes a great respite for people from lower income groups that are unable to afford lavish weddings but have permission to use the *dharamshal* for close family gatherings and wedding receptions. The popularity of Alandi for weddings has increased, and in 2008 an estimated 200 weddings took place in a single month during the festive season of the Hindu calendar (Supekar, 2008). The municipal area of the town is 698 hectares in size, and as of the 2011 census has a resident population of approximately 28,654.

Jejuri

Jejuri is a place for the performance of ritualistic pilgrimage traditions dedicated to Khandoba (or Malhari Martanda) – the presiding deity worshipped as a *Kuldaivat*or family-god mainly by warrior clans, farmers, herders, and priest families in the Deccan region. The mythology of Khandoba, however, cuts across the pan-Hindu pantheon, where more broadly he is seen as the 'Maratha manifestation of Shiva, identified with Kartikeya (standing atop a hill with his consorts), and with Vishnu (Kalki-like with horse and sword) and with the sun (sun-yellow turmeric and the name Martanda)', while at the local level he is believed to have taken the 'daughters of weavers, tailors, horse grooms, [and] soldiers' as his wives from many of the local communities (Patnaik, 2017). The historical prominence of Khandoba is attributed to the royal patronage accorded to the town by the Maratha rulers in the 17th century. The place-stories (*sthala-Puranas*) called the *Malhara Mahatmya* and the *Martanda Vijaya* (composed about 500 years ago by local priests in Sanskrit and Marathi), mention that 'a folk deity cult is gradually assimilated into a grand Brahmanical Hindu superset' (Patnaik, 2017) and thus into a tradition of pilgrimage that involves elaborate rituals for the deity.

Most pilgrims pay homage to Khandoba as part of lifecycle rituals: *kul-dharm*, which means the dharma (or duties) of a person attributed to his lineage as per religious traditions, and *kul-aachaar*, which refers to the rituals that a person should follow as per his family lineage. To perform such rituals, pilgrims are required to seek the services of the religious specialists known as *guravs*, who assist visitors with lodging and boarding in their houses, bring them to the temple, and prepare offerings for Khandoba in return for a fee (*dakshina*). As a social system, Guravs have hereditary rights to serve pilgrims,which rights belong to certain social groups and come from certain geographic areas – a key feature of traditional pilgrimage economy based on religious rituals. At present, almost 90% of residents are engaged in the pilgrimage economy in some way. However, since most of the clientele is comprised of low-income social groups from the region, the scale and volume of the pilgrimage economy is limited. Generally, most people are day visitors, where they come to Jejuri, perform rituals at the temple, and then return home the same day. This has led to an absence of major accommodations in the city. As a case in point, there are only two dharamshal in the town. The temple-complex is managed by the Shri Martand Devasthan Jejuri Devasthan, a charitable trust. The municipality covers an administrative area of about 668 hectares in size and has a population of approximately 12,000 people.

Pandharpur

Pandharpur is a centre of devotional worship dedicated to Vithoba (or Vitthal) – a regional deity believed to be an incarnation of Vishnu (Dhere, 2011). Pandharpur is eulogized by several notable Marathi poet-saints such as Dnyaneshwar, Tukaram, Namdeo, and Chokhamela following the tradition of *bhakti*to Vithoba (Feldhaus, 1986). Besides ritual bathing in the Bhima River, the most significant aspect of worship is a *wari*, where faithful devotees of the poet-saints and Vithoba called *warkaris* walk to Pandharpur. These pilgrimages, where thousands of devotees carry the footprints of their patron saints in palkhi processions to the river, take place on four important occasions – the eleventh lunar day of the bright fortnight of the months of Chaitra, Ashadh, Kartik, and Magh as per the Hindu lunar calendar. Of these, Ashadh Ekadashi is considered the holiest as it marks the day when Vishnu goes to sleep for a period of four months (known as Chaturmas). On this day, devotees undertake fast in the belief that Vishnu will bless them. Participating in Ashadhi wari is meritorious and hence it attracts the largest numbers of pilgrims in terms of scale. During each of these pilgrimages, devotees stay in Pandharpur an average of three to four days (see Mokashi, 1987).

Pandharpur also has an organized pilgrimage economy which developed around the 17th century when the temple received royal patronage from the regional rulers. The hub of religious activity centres on the grand temple housing the idol of the divine couple of Vitthal-Rukmini, the presiding deity. Pilgrims come to *darshan* or to view the idols of the deities Vitthal and Rukmini and engage in worship rituals and services in their honour. These rituals for pilgrims are performed in the temple by priests known as *Badvas*.[1] The ritualised pilgrimage economy ensures that large volumes of visitors visit Pandharpur throughout the year. More than 400 registered *maths* or *mutts* (monasteries and ashrams) belonging to various groups of warkari (sampradaya) provide necessary lodging and boarding facilities for their followers. The municipal council administers an area of 1,717 hectares and has a population of approximately 98,000 as of 2011.

Shegaon

Shegaon is a pilgrim-town dedicated to the worship of Shri Gajanan Maharaj – a saint who resided in Shegaon from 1878 to 1908. Devotees of this guru believe in his miraculous powers to help the needy and that by following his teachings and seeking darshan of his idol at his *Samadhi* (tomb) they can experience a positive change (SGMS, 2007). Here, personal devotional worship is more important than guided ritualistic traditions, leading to a near absence of occupations related to religious service. A private charitable trust called Shri Gajanan Maharaj Sansthan (SGMS) manages the temple complex with strict adherence to the saint's philosophy.[2] The temple complex built around the samadhi (tomb) includes a waiting hall for approximately 10,000 visitors, a dining hall where 5,000 people are served at one time, and other facilities for pilgrims. Using the traditional model of religious charity and patronage, SGMS has constructed six pilgrim-lodges that provide about 700 rooms and community halls that accommodate about 5,000 beds (SGMS, 2007). About 2,000 employees work for the Trust across 42 departments, but most visitor services are provided through an elaborate system of volunteering.

On the outskirts of the town, SGMS has developed a religious theme park called Anandsagar and a residential colony comprising 1,000 apartments (Shinde, 2018). This 300-acre theme-park includes 'a huge water reservoir, an artificial island housing a spiritual center, and more than fifty thousand trees [and] showcases religious artifacts and statues, besides attractions such as a museum, an aquarium, a 3,500-seater amphitheater, and an amusement center' (Shinde, 2018). Because most visitor activity is concentrated within the temple complex and the Anandsagar religious theme park is controlled by the SGMS trust, the town itself has very few lodges and hotels.

While pilgrimage serves as the main economic sector in the town, Shegaon is also the administrative headquarters of the surrounding region, covering many villages outside the main town, leading to the presence of multiple economic sectors including agriculture-related occupations and regional transport (given that it has a railway junction). The town has an area of 840 hectares and a resident population of 61,000 as of 2011.

Shirdi

Over time Shirdi has evolved into a pilgrim-town with a pan-Indian catchment area because of its ties to its patron-saint Sai Baba, a 20th-century guru. The charisma of Sai Baba as a fakir who performed miracles for the poor and needy was instrumental in making him popular during his lifetime; his popularity increased after his demise as followers continued to believe in his divine presence in Shirdi and in his ability to answer their prayers. The phenomenal increase in the followership of Sai Baba has been widely discussed in literature on contemporary religion in India (Kunte, 1976; Sheperd, 1986; Rigopoulos, 1993; Srinivas, 1999; Ghosal and Maity, 2010). The tremendous faith in Sai Baba's divine interventions for followers translates into a very high rate of repeat visitation to Shirdi. Since this type of visitation is beyond the tradition of pilgrimage and involves travel that has more touristic aspects, it is better understood as religious tourism (refer to Shinde, 2011). This form of visitation is also a result of the extensive pilgrimage infrastructure created by the Shri Sai Baba Seva Sansthan Trust (SSST), a public trust that manages the temple complex. The institutional structure of SSST has changed several times since it was first established after Sai Baba's death in 1918, but its main purpose to develop infrastructure for the provision of lodging, boarding, and other public amenities for pilgrims using donations from devotees has continued. SSST has accommodation facilities to host approximately 7,000 people, and this number increases when the congregational halls are used as dormitories (SSST, 2008). Other facilities include a state-of-the-art dining hall with a capacity of 6,000 people and public toilets with a capacity of 1,000 users. SSST also provides a free shuttle bus services for visitors between the accommodation centres, the temple complex, and the dining hall. SSST presently employs more than 4,500 people and has a substantial income (more than $69 million USD in 2016) of which almost half comes as direct offerings from devotees (SSST, 2017). SSST plans to develop additional accommodations for about 15,000 overnight visitors over the next few years.

In Shirdi, visitors come for the *darshan* of the saint's idol and visit places associated with Sai Baba, including his Samadhi, the *gurusthana* or the seat of his guru, Lendi Baug (a garden raised by him), a masjid or mosque known as Dwarkamai (all of which is found within the temple complex), and the *chavdi* (or the office of the village clerk) where he used to sleep. In the temple complex, a waiting hall holds up to 10,000 visitors and where movement is regulated through multiple queues. This helps in managing the crowds and controlled entry into the inner sanctum or centre of the shrine where the main idol of Sai Baba is installed over his *Samadhi*. This area forms the sacred core of the town where intense pilgrim flows are observed. Around this sacred core are approximately 2,000 commercial establishments, including 200 guesthouses for lodging, more than 150 private homes that rent out rooms, and between 150 and 200 permanent and temporary restaurants and shops selling Sai Baba novelties, trinkets, souvenirs, and prasad (i.e. food offerings that are consecrated by presenting them to the deity), and materials required for temple worship, such as garlands and sweets (Shinde, 2017). There are also tour operators that offer package tours to other religious places in the vicinity, such as Shani Shingnapur (a site dedicated to deity of Shani), Nashik (the site of the Kumbha Mela), and Tryambakeshwar (a pilgrim town dedicated to Hindu god Shiva). This has led to Shirdi becoming a multi-dimensional religious tourism destination. The municipal council of Shirdi covers an area of 1,298 hectares and has a resident population of approximately 34,000.

Tuljapur

Tuljapur is the home of Devi Tulja Bhavani, a regional incarnation of the Goddess of Power Shakti, which several clans within Maratha society consider as their family deity (*kuldaivat*) (Pohnerkar, 1961; Prayag, 1977). Tulja Bhavani is a goddess who protects families, grants boons, and responds to vows. Because of this, lifecycle rites are dedicated in her name and visits to her temple are important requirement for devotees. As such, pilgrims engage in elaborate rituals and give offerings to seek her blessings. Historical sources suggest that ritualized pilgrimages to Tulja Bhavani and her temple gained prominence during the Maratha rule after the Maratha King Shivaji sought blessings of the goddess in his quest to build a Hindu kingdom. Several hegemonic lineages of priests from the 17th century have maintained records of visits by royal patrons and people from the region in their official registers (Dhere, 2008; Shinde, 2011).

Pilgrims visit the temple complex perched atop the mountain, several shrines in the surrounding area, and a lake where they must undertake a ritual bath. The pilgrims engage in elaborate rituals where Devi is worshipped in her human form but represented through an idol. As such, every day the idol of the goddess is bathed, clothed, fed four meals, entertained with devotional music, and put to bed at night (Pohnerkar, 1961). Between these daily events, pilgrims make their offerings to the Devi, which range from a plate containing a piece of cloth to a coconut, a bowl of puffed rice, a flower garland, and sweetmeat, to a four-course meal. To accomplish this, pilgrims need to seek the services of religious specialists known as *Pujaris* who are pulled from three distinct groups defined by hereditary traditions and provide all ritual services necessary for pilgrims in the temple and lodging and boarding in the town.[3] Approximately 70% of the residents are engaged in religious occupations, and most pujari families own shops where devotees can buy devotional items and offerings, including religious trinkets, *prasad*, and other articles necessary for performative rituals. The pilgrimage economy in Tuljipur is driven by rituals and offerings (Shinde, 2011).

The main temple of the devi is administered by a public charitable trust under the chairmanship of the District Collector and is comprised of representatives from the lineages of the temple-pujaris. The administrative area of Tuljiapur is 416.44 hectares and has a resident population of approximately 35,000.

Visitation to the Six Destinations

In general, visitor numbers are quite high in the six pilgrim-towns and tend to swell on weekends and on days that are considered auspicious for the worship of the presiding deity or saint (Fig. 3.3). For instance, Ekadashi of the dark half of each month (in all there are 24 Ekadashis in Hindu lunar calendar) is hailed by warkaris for the worship of Vithoba, which means that both Alandi and Pandharpur witness peaks on those days (MPCB, 2005). Physical proximity to a metropolitan city such as Pune (population of 4 million) is a major factor in increasing visitation, as seen in case of Alandi (16 kilometres to the north) and Jejuri (50 kilometres to the east). Festivals also bring in huge crowds. For instance, *Navratra*, a nine-day festival dedicated to Devi, sees more than 500,000 visitors in Tuljapur (Shinde, 2011). This visitation adds to the environmental impacts at these six case study sites, as discussed below.

Environmental Challenges

Most studies separate the environmental impacts of visitors in destinations as either 'direct'or 'indirect' (e.g. Cohen, 1978; Butler, 2000; Shinde,

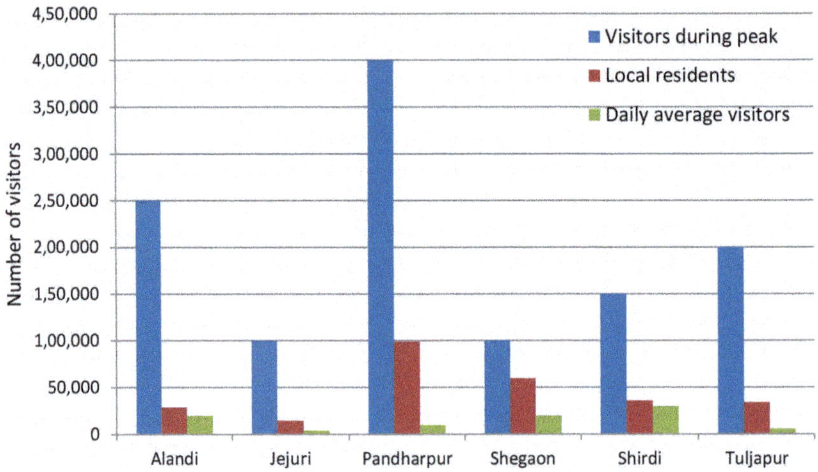

Fig. 3.3. A chart showing the visitor flows in the six destination case study sites, circa 2010 (author).

2007; Terzidou *et al.*, 2008; Alipour *et al.*, 2017). Direct environmental impacts are those that are caused directly due to the presence and actions of visitors, such as congestion, the consumption of resources, the generation of waste, and wear and tear of religious objects and establishments. Indirect environmental impacts are not necessarily caused by visitors themselves but by the services and changes in an area due to tourism more generally, such as the development of tourism infrastructure, economic restructuring, urban growth, real estate development, land use changes, and cultural impacts (Ambrósio, 2003; Shinde, 2017).

In all six towns, the visitation rates during religious festivals is several times more than the number of daily visitors. However, while such visitor volume is observed for a maximum of a day or two, this influx of pilgrims, religious tourists, and tourists more generally causes serious direct impacts. Several studies conducted by the Maharashtra Pollution Control Board during and after the annual pilgrimage confirm huge quantities of 'biodegradable [raw sewage] waste entering in the river', worsening the situation of 'sanitation and sanitation and hygienic conditions' in the town of Pandharpur (MPCB, 2005).

The size and scale of the towns has some bearing on the types and magnitude of the environmental problems they experience. In smaller towns such as Shirdi and Alandi, the ratio of visitors to residents is very high, whereas in larger towns, such as Pandharpur, the ratio of visitors to residents is relatively lower and hence any impacts from visitors are more easily assimilated into the existing system of infrastructural and social services. However, due to its cumulative nature, both direct and indirect impacts from pilgrimage and religious tourism still contribute to the overall environmental degradation of these towns.

A large proportion of visitors to all towns are day-trippers, and while their stay is short, their presence and actions do cause some impact, however minimal that impact may be. The direct impacts are more visible in places such as Tuljapur and Jejuri where life cycle rituals, which revolve around consumptive practices, involve food and other types of offerings. As noted by the MPCB (2005, p. 11),

> The daily flow of pilgrims also creates unhygienic conditions around Pandharpur. The hotel food is usually consumed by the visitors and the waste is thrown into the surface drain. The pilgrims offer flowers, garlands, coconuts, etc. in the temple and the river. The waste generated by these offering remains undisposed and untreated. In course of time decomposition takes place generating a foul polluting [of] the environment.

However, in the case of Shirdi and Shegaon, where devotional worship is simply *darshan*, the generation of waste tends to be limited.

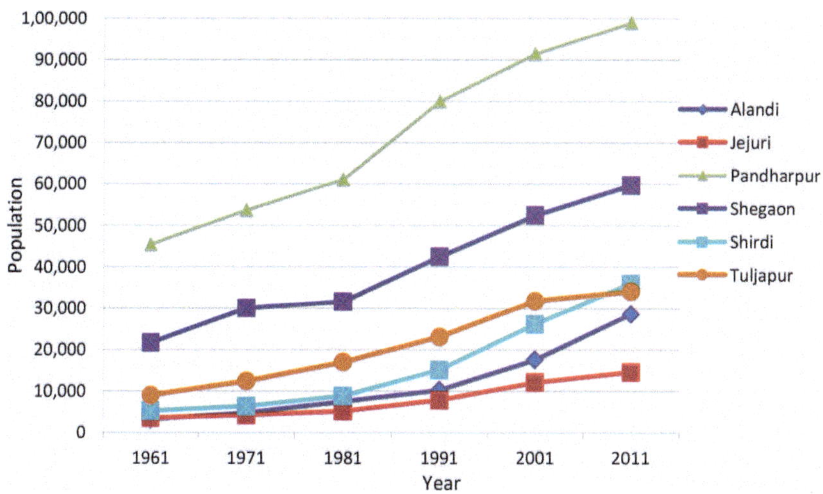

Fig. 3.4. A graph showing the population growth in the six case study destinations (1961–2011) (author).

However, their limited impacts may be offset by larger numbers and frequent visitation rates.

The indirect impacts of pilgrimage and religious tourism are far-reaching and long-term, and therefore more concerning from an environmental sustainability standpoint. Increasing numbers of visitors drives the development of tourist infrastructure (e.g. food, transportation, and accommodation). Indirectly, this also means that more people come to these towns in search of livelihood opportunities. As shown in Fig. 3.4, there has been considerable growth in the resident population in the six destinations over the last 50 years, which has led to rapid urbanization and its attendant developmental issues.

Environmental impacts are more pronounced in places where natural resources are associated with the pilgrimage economy. For instance, water pollution has emerged as a serious problem in Alandi and Pandharpur owing to the increase in sewage discharge into the nearby river. Deforestation and soil erosion are major challenges in the mountain-top sites of Jejuri and Tuljapur. Also, in all six towns, the erstwhile agriculture-dominated rural landscape is rapidly being changed due to its reconstitution into urban expansion processes. As Shinde (2017) notes,

> The outlying areas which were once used for agriculture and cultivation are converted into large plots and most are earmarked for construction of hotels ... This is also the prime area for real estate development: the lifelong devotees choose to have a permanent accommodation in Shirdi either in forms of apartments or bungalows. This demand driven by religious needs drives real estate business and because devotees are ready to pay price for emotional reasons, the property prices in Shirdi are quite high– similar to what one would find in major cities like Pune and Nashik.

The relationship between pilgrimage, religious tourism, and urban change can also be illustrated in the case of Shirdi, where the resident population has grown at a phenomenal rate of almost 8% a year due to rapid and exponential growth in visitation in recent years. Around the temple complex in Shirdi, congested commercial areas have developed with a high concentration of shops, hotels, and restaurants, many of which are illegally built. Similar conversions of agricultural lands to urban uses are evident in all other sites except for Jejuri, which has a lower rate of visitation. The change in land use is not only driven by real estate speculation but also the need to increase ritual and religious services. In Shirdi, the local population 'used to cultivate cereals like jowari (millet), wheat[,] etc. [,] but now they [have] switched over to flower cultivation in their agricultural fields which is very much in demand for offering of pujas in the Sai Baba temple' (Ghosal and Maity, 2010).

The environment of the case study towns is also changing in terms of its built heritage. In the past, royal patronage not only helped in providing services for pilgrims, but also led to the building of grand buildings and other structures such as riverside *ghats* (embankments and flights of steps along river banks to provide access) and *kunds* (water-tanks) in most pilgrim-towns. For example, there used to be more than 15 elaborately carved water tanks in Tuljapur. However, due to lack of investment they have been left in a dilapidated condition (Prayag, 1977). Also, the diversity of different forms of accommodation and housing has produced a rich sociospatial fabric in the towns. However, in Pandharpur and Tuljapur, simpler accommodations are now under threat due to the rapid pace of construction of modern hotels and amenities on many sites, destroying this religious heritage in the name of modernity and progress. The excessive construction of urban structures has brought about a change in the character of these towns, once regarded as religious sites, and now considered to be urban places with a religious attraction (Shinde, 2017).

All the six towns have evolved rapidly from smaller villages to larger towns, and without the appropriate investment from local and regional governments have only developed limited physical infrastructures with low carrying capacities. Most have a pressing need for augmenting environmental services, including investments in their water supply, sanitation, and solid waste infrastructure, and management to cater to their growing populations. For instance, in Alandi there is no 'proper underground drainage system and at present the storm water drains carry the sullage as well as sewage water [and] all the sewerage is let into the Indrayani River without any treatment' (Wilbur Smith Associates, 2005). The inadequacy of basic infrastructural capacities is further compounded by a floating population of pilgrims. At the same time, most of the towns are small, and it would be difficult to scale up environmental services due to the spatial organization of their urban situations.

There are also intangible processes through which indirect changes become apparent. These processes emanate in both the religious and cultural domains. For instance, religious service providers use hereditary hegemonic positioning and lay claim to the necessary religious capital to have responsibilities to develop infrastructure for religious tourism. But this process can best be termed informal, as these actors tend to exclude themselves from formal systems of planning and development, leading to, in the case of Shirdi (Shinde, 2017), ad-hoc development without any architectural merit.

Institutional Responses to Environmental Degradation

Pilgrim-towns generally received royal patronage through which they were governed in the precolonial period. Another layer of administration was introduced by the colonial government as it viewed such towns that were religious gathering places as being prone to diseases and epidemics (Freitag, 1989; Lochtefeld, 2010). Even though these were small compared to other major urban centres, such places had municipalities through which the government would provide temporary resources, such as police and sanitation infrastructure to maintain hygienic conditions during religious fairs and festivals. Such a system of responding to environmental issues related to pilgrimage events has continued in the post-independence period (Maclean, 2003; Ruback *et al.*, 2008). For instance, during the four main pilgrimages in Pandharpur, large amounts of water are released from the Ujani Dam located upstream so that human waste can be washed downstream. In addition, local and regional governments provide temporary toilets along the riverbank (MPCB, 2005). One interviewee in Jejuri told the author that 'Jejuri knows it's routine ... we deal with 12–15 small [and] big pilgrimages annually ... the District Collector [the administrative officer for the district] provides about 150 temporary latrines and pit latrines.'

The dramatic growth of pilgrim-towns, however, has challenged the ability of these small municipalities to deal with the rising environmental impacts of pilgrimage and religious tourism. Because of the small scale of their operations, they have considerable constraints, limited revenues and a lack of human resources and technical capacity to provide environmental services. For example, of the total staff of 114 at the Tuljapur municipality, only one is an engineer. Also, only 5% of its annual budget (INR 1.3 million = approximately $18,800 USD) is allotted for the management of solid waste, which is a meagre amount compared to the necessary expenditure required to manage the huge quantities of waste generated in the town. This leads to a perpetual problem of waste accumulating over time, which then snowballs into several public health crises.

According to government policy, issues related to urban growth and land use are to be addressed through development plans, which include a proposed land use plan to allocate land for zoning and to develop control regulations that address future housing and infrastructure needs based on projected demographic growth and assumptions about present and future economic drivers of growth for the next 20 years (Das, 1981). A close analysis of development plans for the six case study destinations reveals some common trends. The first trend is that these plans do not consider the pilgrimage economy, which is a major driver for economic growth in these towns, and therefore the plans were based on incorrect assumptions leading to failed population projections. For example, the revised development plan for Tuljapur acknowledged that the 'projected public amenities [,] facilities, [and] residential zone fall ... short in the sanctioned Development Plan as compared to [the] actual [growth] in population [between]1995 and 2005' (Town Planning Department, 2008). The disconnect between the proposals in the development plans and present realities was a common occurrence in the development plans of all six towns.

The second trend was that projections in terms of future types of land use and the amount of land needed for those uses did not correspond to the actual use of land and its distribution. For example, much of the designated residential area in Tuljapur is presently used for commercial activities related to the pilgrimage economy, while in Pandharpur, 'about 39% more area than what was proposed for residential use in [the] sanctioned plan is actually under residential use' (Town Planning Department, 2008). A third common trend in the development plans was the problem with the policy of 'land reservations' for providing facilities in future (as per the Town Planning Act, the government has the authority to designate some land as reserved for public purpose which it may acquire at a later stage). This policy has proven to be misplaced, as its implementation is fraught with many problems, the most significant being the inability of resource-strapped municipalities to acquire land. Also, these land acquisitions are not always used for their intended purposes. For example, in Tuljapur, implementation was less than 12%, with only 11 out of 89 'reserved sites' developed. Of these sites, only five had an explicit public purpose while the rest were for commercial use that could generate revenue for the municipality. A similar story occurs with proposals for the widening of roads to address congestion. Such proposals are generally resisted by local communities because their businesses are these along the main access roads to temples, and these businesses would need to be removed to accommodate this road-building. In fact, during public consultations for the revised 2002 development plan for Pandharpur, 1,670 objections were raised against the proposals for the widening of roads in the town (Town Planning Department, 2002).

Another trend was that in all the development plans the issue of increases in visitation during religious holidays and festivals was acknowledged, albeit only in passing, since the mandate of the development plans was to provide for the basic amenities for resident populations. The revised development plan for Shirdi, for example, acknowledged that 'the magnitude of increase in pilgrims itself is a very serious problem which is required to be attended to while formulating [the] Development Plan'. However, a floating population of only 15,000 was considered for the purposes of this planning document, and a proposal was made for reserving 15 hectares of land under 'public and semi-public use' for accommodating the needs of these visitors (Town Planning Department, 1992) — a grossly inadequate amount of land in view of more than 35,000 daily visitors.

Recognizing the limited capacity of local municipalities, since the early 1990s the Ministry of Urban Development for the state has formulated 'special development plans' to help with the development of infrastructure and enhancement of environmental services in pilgrim-towns using an ad-hoc approach. So far, all six towns have received some sort of funding. For instance, in 2009 Pandharpur received funding for the construction of an underground sewer system and public toilets, the improvement of roads, electrical generation, and updating the water supply distribution systems. The top-down approach implemented for these grand projects was fraught with many problems. For example, these are long-term projects that require considerable land acquisition and therefore were met with resistance from the local community, particularly those whose daily livelihoods were most affected. Also, no maintenance funds were given to the local municipalities, so they were not able to service the infrastructure after their development. Also, there was the problem of underestimating the costs of these projects. In Shegaon, for instance, the cost of implementing the plan (called Shegaon Vikas Arakhada in Marathi) for the development of the town soared to three times its initial estimate (from INR 9.2 million to INR 25 million), leaving many of the works incomplete. In Tuljapur, the only visible outcome was the widening of roads and a block of public toilets which are seldom used by visitors. Overall, such projects delivered some infrastructure, however in many cases they were directed by elected representatives for the resident population, leaving the needs of visitors unaddressed.

A more recent intervention in pilgrim-towns is the formation of Urban Development Authorities through special legislation by the state. These urban authorities are mandated to prepare a structural and regional plan for a much larger area surrounding the pilgrim-towns. These quasi-state institutions have a board-like structure comprised of 14 members from different state government departments and representatives from electoral constituencies. These authorities make decisions for the pilgrim-towns under the leadership of the Collector of the District. However, there are concerns over inefficiencies and the lack of effectiveness of these authorities. For instance, the Tuljapur Urban Development Authority (TUDA), which was constituted in 2008, had only two representatives from the town. The TUDA proposed a development plan which is not a strategic plan per se, but rather a compilation of activities focused around three areas: road works, the construction of buildings, and beautification projects, with the road works projects taking up more than 75% of the budget. In this capital-intensive plan, little consideration was given to visitors and their needs, with less than 1% of the budget being allocated for the building of new toilets. Even through the TUDA develops these plans, the responsibility for their implementation is thrust on local agencies that are already burdened and lagging with regards to service provision. As such, there are bureaucratic delays between the decision-making and implementation of development plans. Moreover, half of the budget is the responsibility of the municipality, which in most cases has no technical capacity or expertise to implement these engineering-heavy projects.

In the context of tourism, the only responsible agency in the region is the state-level Maharashtra Tourism Development Corporation (MTDC). Recognizing the tourism potential of religious sites, the MTDC has actively been promoting religious tourism through its policies and campaigns. However, the Corporation has no jurisdiction over the actual places regarding infrastructural improvements. Moreover, the MTDC requires that private sector entities should be registered with the government to become 'eligible units' for tax concessions and other incentives (Department of Tourism, 2006). For example, a firm interested in constructing a motel that includes public amenities for visitors would not be required to pay tax on their income for the first five years as this would help in promoting tourist facilities in those areas. This policy and corporate focus exclude many pilgrimage and religious tourism service providers that operate through social and informal business networks, leading to difficulties in addressing and regulating problems related to this informal pilgrimage economy. In response to this informality of providing pilgrimage services, government officials in Shridi removed more than 200 shops that were 'deemed illegal' from the vicinity of the main temple complex.

It is also a challenge to find community-based organizations that are interested in responding to the above-mentioned environmental challenges because of the socioeconomic and cultural fabric wherein most people are engaged in religious- or tourism-related occupations. An informant mentioned to the author that because of the ease of finding jobs within the pilgrimage economy, less than 60% of the young people of Jejuri complete their secondary education. These lower education levels in turn contribute to a general apathy about fixing environmental issues in the town. In addition, and probably more pressing, the general sentiment among locals in all the towns is that their presiding deity, as their guardian, will take care of the environmental issues in their town (see

Chapter 9, this volume). Residents seem to attribute everything to the divine, and see religious and cultural service as paramount in their lives rather than articulating and addressing these environmental problems.

In sum, institutional responses to environmental problems in the six towns are fraught with difficulties and challenges. Pilgrimage events are fairly well managed, but these pilgrim-towns struggle with the cumulative impacts of both the high daily volume of visitors and the peak visitation rates during religious festivals and holidays that cause long-term changes to population growth and urbanization. There is also the systemic problem of planning, where religious heritage resources are often under the control of temple trusts which have the mandate to manage the temple complexes and regulate flows to ensure that visitors can perform the necessary religious rituals. However, as seen in Shirdi and Shegaon, these temple organizations are bound by their jurisdictions and hardly address environmental problems outside of the sacred site. In towns with informal pilgrimage economies, the focus on religious rituals and performances dominates environmental concerns, leading to the neglect of the physical environment.

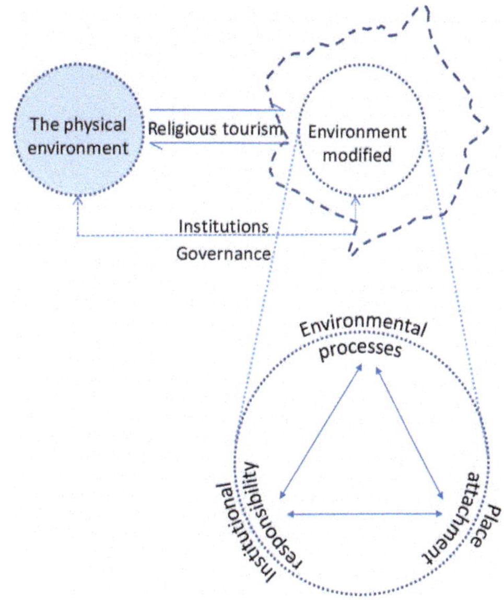

Fig. 3.5. A proposed conceptual framework for environmental analysis in a pilgrimage place.

A Conceptual Framework for Environmental Analysis

The comparative analysis above highlights that the environmental challenges and responses in the six pilgrim-towns share many commonalities and only minor differences. From this analysis a conceptual framework can be developed to better understand environmental issues and environmental management in religious tourism destinations. This framework is shown in Fig. 3.5 and builds on the conceptual approach proposed by Shinde and Olsen in Chapter 1 to explain the relation between religious tourism and the environment in a sacred place (pilgrim-town).

The environment in pilgrim-towns is modified in many ways due to religious tourism. To make sense of the modified environment, the framework proposed in Fig. 3.5 is oriented around three interrelated analytical categories: *environmental processes, institutional responsibility,* and *place attachment.*

A focus on *environmental processes* emphasizes the scale, scope, and geographic area within which environmental impacts are caused by pilgrimage and religious tourism in explaining causal relations and influences. A comparison of Shirdi and Jejuri helps to illustrate this. In Shirdi, the scale of environmental impacts is large in terms of the magnitude and growth of its residents and visitors, which growth leads to consequent environmental impacts. In terms of scope, the environment impacts range from regular on-road traffic congestion to large-scale urban land use conversions that add to this congestion. In terms of geographical area, the temple trust is active in the sacred core of Shirdi, but the public domain of the town is severely constrained. In Jejuri, the scale of environmental impacts is relatively small owing to the limited area of visitation and pilgrim activity, the low base of visitors (narrow regional catchment) and the lower growth rate of visitors. The scope of environmental concerns is narrower and related to the limited duration of stay of visitors and their focus on performing rituals. The geography of Jejuri constitutes the spatial manifestation of the traditional pilgrimage economy, which means more involvement by the local community and less direct involvement from the temple trust.

Institutional responsibility includes collective ways of articulating and defining environmental problems and addressing them through moral, ethical, and legal authority. As institutions, local municipalities have all three kinds of responsibilities towards residents, but not necessarily to visitors, considering that they are not a part of the electoral constituency. This means that problems related to wider and long-term environmental concerns caused by visitors are not likely to be the top priority of temple trusts. As custodians of the sacred place, religious institutions may have moral authority and a legal mandate to serve the temples and visitors, but legally have no influence in articulating or influencing the priorities of town governments. Institutions at the state level do play a role in the larger institutional development of the region through the Urban Development Authorities in pilgrim towns. The examination of authority and responsibility along these lines may offer better insights regarding environmental (in) action in pilgrim-towns.

Place attachment is crucial to understand different motivations for institutions, groups, and individuals to participation in solving environmental concerns. The unwavering faith in town deities by town residents contributes to a spirit or sense of place, which spirit is further reinforced through traditions, rituals, festivals, and other processes that sanctify the town. The sacred nature of these towns is also manifest in the historical development of exquisite temples, riverside *ghats*, and the mansions of elite patrons, as well as the ordinary houses of priests that comprise part of the religious infrastructure and sacred geography of the town. However, as changes in the urban landscape reflect efforts to accommodate growing numbers of pilgrims and religious tourists, this commercialization of the sacred alters the spirit of place from a religious centre to an urban city. Indeed, in many cases, any concerns regarding the degradation of the environment come from changes to a town's religious heritage and sense of place. The intangible aspects of place attachment, such as this unwavering belief in the miraculous powers of the deity to protect and maintain the towns and the need to carry on religious and cultural ways of life, often lead to inaction on the part of religious actors. A deeper investigation into place attachment will unpack what different stakeholders feel about these pilgrim-towns and their problems, and why they choose to address or not address them.

This framework of 'process, institutions, and place' seems to hold a comprehensive and explanatory rigour to better explain environmental management in religious tourism destinations, which rigour will need to be tested by future research.

Conclusion

The aim of this chapter was to present a comparative study of pilgrim-towns in the state of Maharashtra in western India to help develop a conceptual framework to explain the complexities of managing environmental issues in pilgrim-towns. The similarities in environmental challenges found in the six pilgrim-towns illustrate that this comparative approach can be fruitfully used to arrive at a conceptual framework. Furthermore, despite an overarching responsibility for dealing with the environmental impacts of pilgrimage and religious tourism, the role of formal government institutions ranged from being limited and constrained – where the non-state actors and other factors dominate – to complete control of the environment in which pilgrimage and religious tourism operates. Although the pilgrim-towns studied are from one state in India and belong to the Hindu religion, the findings are indicative of similar environmental challenges and management measures that operate in pilgrimage and religious tourism destinations across the globe (Shackley, 2001; Henderson, 2011; UNWTO, 2011).

Pilgrim-towns, as pilgrimage and religious tourism destinations, are heavily imbued with sacred values and religious meanings that shape the worldviews of residents and visitors (Terzidou *et al.*, 2008; Alipour *et al.*, 2017) that further lead to either action or inaction regarding the environmental impacts of religious activities in such places. This intangible value is manifest through creating tangible representations of religious paraphernalia and sociocultural interactions between residents, religious workers and leaders, and visitors. The process of sanctification that occurs through the reinforcement of the sacred nature of these pilgrim-towns through continued pilgrim activity (see Bremer, 2004, p. 307) continues to attract people and sets in motion several kinds of environmental processes that have the potential to alter the physical and metaphysical environment of these pilgrim-towns. Different types of institutions attempt to articulate and respond to environmental problems, but are constrained by their responsibilities, policies, and the lack of money and authority. These core ideas are interweaved into the proposed conceptual framework of *processes*, *institutions*, and *place*.

In this study of six pilgrim-towns, the focus on the broader patterns of environmental impacts took precedence over details, particularly with regards to the nuanced sociocultural aspects of the pilgrimage economy at each site. For example, environmental impacts on specific natural features at these sites, such as rivers and forests, require a more in-depth analysis than what has been included here, and as such can form a major field of research on its own. The focus on institutional responses meant that individual and collective behaviours towards the environment and environmental ethics was not dealt with in adequate detail in this chapter. This field of cognition and perception has great value in terms of the prevalence of certain environmental discourses in pilgrim-towns and is a topic that should be explored further. Overall, this chapter offers a fresh perspective and a significant conceptual contribution in expanding the scope of environmental management studies in pilgrimage and religious tourism destinations.

Notes

[1] There are also other religious specialists in the town. The ones involved in the maintenance of the temple and providing services during festivals and processions are called *sevadharis* or 'the servants of the god'. There are also several gurus and saints who are visited by pilgrims for spiritual counsel and as part of their wari.

[2] Sant Gajanan Maharaj practiced the philosophy of service to the poor and helpless. He also emphasized Adhyatmic (spiritual) enlightenment, improvement of health and education to train the mind, body and brain as the goals for the sansthan that he established.

[3] One group of pujaris include sixteen family lineages who holds the exclusive rights to perform the worship rituals in the temple for patrons. The second and much larger group of pujaris has over 3000 families which have the hereditary rights to provide lodging and boarding services to pilgrims, assist them in preparing food offerings for the Devi, and arrange worship ceremonies. The third group is required to offer specialized temple services, including cleaning, maintenance, lighting of lamps, and helping during the processions. With expansion of clans and families over generations, such hereditary rights have become entangled and complicated leading to prolonged litigations over the rights over and income from the temple.

References

Alipour, H., Olya, H.G.T. and Forouzan, I. (2017) Environmental impacts of mass religious tourism: from residents' perspectives. *Tourism Analysis* 22(2), 167–183. DOI: 10.3727/108354217X14888192562285.

Alley, K.D. (2002) *On the Banks of the Ganga: When Waste Water Meets a Sacred River*. University of Michigan Press, Ann Arbor.

Ambrósio, V. (2003) Religious tourism: territorial impacts in the EU Marian sanctuary towns. In: Fernandes, C., Mcgettigan, F. and Edwards, J. (eds) *The 1st Expert Meeting of ATLAS—Religious Tourism And Pilgrimage Research Group*. Tourism Board of Leiria, Fatima, pp. 43–56.

Bhardwaj, S.M. (1973) *Hindu Places of Pilgrimage in India*. University of California Press, Berkeley, California.

Bhardwaj, S.M. (1997) Geography and pilgrimage: a review. In: Morinis, A. and Stoddard, R. (eds) *Sacred Places, Sacred Spaces: The Geography of Pilgrimages*. Louisiana State University Press, Baton Rouge, Louisiana, pp. 1–24.

Bremer, T.S. (2004) *Blessed with Tourists: The Borderlands of Religion and Tourism in San Antonio*. University of North Carolina Press, Chapel Hill and London.

Butler, R.W. (2000) Tourism and the environment: a geographical perspective. *Tourism Geographies* 2(3), 337–358. DOI: 10.1080/14616680050082553.
Caplan, A. (1982) Pilgrims and priests as links between a sacred center and the Hindu culture region: Prayag's Magh Mela pilgrimage (Allahbad, India). Ph.D. thesis, The University of Michigan.
Cohen, E. (1978) The impact of tourism on the physical environment. *Annals of Tourism Research* 5(2), 215–237. DOI: 10.1016/0160-7383(78)90221-9.
Collins-Kreiner, N. (2010) The geography of pilgrimage and tourism: transformations and implications for applied geography. *Applied Geography* 30(1), 153–164. DOI: 10.1016/j.apgeog.2009.02.001.
Das, B. (1981) Urban planning in India. *Social Scientist* 9(12), 53–67. DOI: 10.2307/3517133.
Department of Tourism (2006) *Tourism Policy of Maharashtra*. Department of Tourism and Cultural Affairs (DoTCA), G.o.M., Mumbai.
Dhere, R.C. (2008) *Shri Tuljabhavani*. Padmagandha Prakashan, Pune, India.
Dhere, R.C. (2011) *The rise of a folk God: Vitthal of Pandharpur*. Oxford University Press, New York.
Dubey, D.P. (1995) *Pilgrimage Studies: Sacred Places, Sacred Traditions*. Society of Pilgrimage Studies, Allahabad, India.
Feldhaus, A. (1986) Maharashtra as a holy land: a sectarian tradition. *Bulletin of the School of Oriental and African Studies* 49(3), 532–548. DOI: 10.1017/S0041977X00045080.
Freitag, S.B. (1989) *Culture and Power in Banaras: Community, Performance, and Environment, 1800-1980*. University of California Press, Berkeley, California.
Ghosal, S. and Maity, T. (2010) Development and sustenance of Shirdi as a centre for religious tourism in India. In: Singh, R.P.B. (ed.) *Holy Places and Pilgrimages: Essays on India*. Shubhi Publications, New Delhi, India, pp. 159–173.
Gladstone, D. (2005) *From Pilgrimage to Package Tour: Travel and Tourism in the Third World*. Routledge, New York.
Haberman, D.L. (2006) *River of love in an age of pollution: the Yamuna River of Northern India*. University of California Press, Berkeley, California..
Henderson, J.C. (2011) Religious tourism and its management: the hajj in Saudi Arabia. *International Journal of Tourism Research* 13(6), 541–552. DOI: 10.1002/jtr.825.
Joseph, C.A. (1994) Temples, tourists and the politics of exclusion: the articulation of sacred space at the Hindu pilgrimage centre of Pushkar, India. Ph.D. Thesis, University of Rochester, Rochester, New York.
Kunte, B.G. (1976) *Ahmadnagar District Gazetteer*. Government Central Press, Bombay, India.
Lochtefeld, J. (2010) *God's gateway: Identity and meaning in a Hindu pilgrimage place*. Oxford University Press, New York.
Maclean, K.K. (2003) Power and pilgrimage: the Kumbha Mela in Allahabad, 1765-1954. Ph.D. thesis, La Trobe University, Melbourne, Australia.
Mokashi, D.B. (1987) *Palkhi: an Indian pilgrimage*. SUNY Press, Albany, New York.
MPCB (2005) *Environmental Status Report of Pandharpur (During Ashadhi Ekadashi).Regional Office*. Maharashtra Pollution Board Office, Pune, India.
Olsen, D.H. (2003) Heritage, tourism, and the commodification of religion. *Tourism Recreation Research* 28(3), 99–104. DOI: 10.1080/02508281.2003.11081422.
Olsen, D.H. (2006) Management issues for religious heritage attractions, *Tourism, Religion and Spiritual Journeys*, Routledge, London and New York, 120–134.
Patnaik, N. (2006) *Sacred geography of Puri: Structure and organisation, and cultural role of a pilgrim centre*. Kalpaz Publications, Delhi, India.
Patnaik, D. (2017) Jejuri: shower of Turmeric. *Pune Mirror* 8/01/2017.
Pohnerkar, N.S. (1961) *Maharashtra chi mata shri Tuljabhavani*. Archaeology Department, Maharashtra State, Aurangabad, India.
Prayag, K.B. (1977) *Shri kshetra Tuljapur Mahatmya*. Pratipchandra Kamlakar Prayag, Pune, India.
Raj, R. and Morpeth, N.D. (eds) (2007) *Religious Tourism and Pilgrimage Festivals Management: An International Perspective*. CAB International, Wallingford, UK.
Rigopoulos, A. (1993) *The Life and Teachings of Sai Baba of Shirdi*. State University of New York Press, Albany, New York.
Ruback, R.B., Pandey, J. and Kohli, N. (2008) Evaluations of a sacred place: Role and religious belief at the *Magh Mela*. *Journal of Environmental Psychology* 28(2), 174–184. DOI: 10.1016/j.jenvp.2007.10.007.
SGMS (2007) *Shri Gajanan Prarthana Strot*. Shri Gajanan Maharaj Sansthan (SGMS), Shegaon, Maharashtra, India.

Shackley, M. (2001) *Managing Sacred Sites: Service Provision and Visitor Experience*. Continuum, London.
Shackley, M. (2004) Tourist consumption of sacred landscapes space, time and vision. *Tourism Recreation Research* 29(1), 67–73. DOI: 10.1080/02508281.2004.11081432.
Sheperd, K. (1986) *Gurus Rediscovered: Biographies of Sai Baba of Shirdi and Upasni Maharaj of Sakori*. Anthropographia, Cambridge, UK.
Shinde, K. (2007) Pilgrimage and the environment: challenges in a pilgrimage centre. *Current Issues in Tourism* 10(4), 343–365. DOI: 10.2167/cit259.0.
Shinde, K.A. (2008) The environment of pilgrimage in the sacred site of Vrindavan, India. Ph.D. Thesis, Monash University, Melbourne, Australia.
Shinde, K.A. (2011) Placing communitas: Spatiality and ritual performances in Indian religious tourism. *Tourism* 59(3), 335–352.
Shinde, K.A. (2012a) Place-making and environmental change in a Hindu pilgrimage site in India. *Geoforum* 43(1), 116–127. DOI: 10.1016/j.geoforum.2011.07.014.
Shinde, K. (2012b) Policy, planning, and management for religious tourism in Indian pilgrimage sites. *Journal of Policy Research in Tourism, Leisure and Events* 4(3), 277–301. DOI: 10.1080/19407963.2012.726107.
Shinde, K.A. (2017) Planning for urbanization in religious tourism destinations: insights from Shirdi, India. *Planning Practice & Research* 32(2), 132–151. DOI: 10.1080/02697459.2016.1198197.
Shinde, K.A. (2018) Augmenting pilgrimages: a religious theme Park in Shegaon, Maharashtra. In: Pinkney, A.M. and Whalen-Bridge, J. (eds) *Religious Journeys in India: Pilgrims, Tourists, and Travelers*. University of New York Press, Albany, New York, pp. 65–90.
Singh, S. (2004) Religion, heritage and travel: case references from the Indian Himalayas. *Current Issues in Tourism* 7(1), 44–65. DOI: 10.1080/13683500408667972.
Srinivas, S. (1999) The Brahmin and the fakir: suburban religiosity in the cult of Shirdi Sai BabA. *Journal of Contemporary Religion* 14(2), 245–261. DOI: 10.1080/13537909908580865.
SSST (2008) *Sai-lila: Annual Report of the Shree Shirdi Sai Sansthan Trust*. Shree Shirdi Sai Sansthan Trust, Shirdi.
SSST (2017) *Sai-lila: Annual Report of the Shree Shirdi Sai Sansthan Trust*. Shree Shirdi Sai Sansthan Trust, Shirdi.
Stausberg, M. (2012) *Religion and Tourism: Crossroads, Destinations and Encounters*. Routledge, Abingdon, UK.
Supekar, N.R. (2008) *Environmental Aspects in Development Plan of Religious Towns, Case Study - "Alandi"*. Masters Thesis. College of Engineering Pune University, Pune, India.
Terzidou, M., Stylidis, D. and Szivas, E.M. (2008) Residents' perceptions of religious tourism and its socio-economic impacts on the island of Tinos. *Tourism and Hospitality Planning & Development* 5(2), 113–129. DOI: 10.1080/14790530802252784.
Timothy, D.J. and Olsen, D.H. (2006) *Tourism, Religion and Spiritual Journeys*. Routledge, London and New York.
Town Planning Department (1992) *Development Plan for Shirdi 1991-2011*. Town Planning Department Government of Maharashtra, Shirdi, India.
Town Planning Department (2002) *Draft Report of the Second Revision to the Development Plan for Pandharpur*. Town Planning Department Government of Maharashtra, Pandharpur, India.
Town Planning Department (2008) *Draft Report of the Revised Development Plan for Tuljapur*. Town Planning Department Government of Maharashtra, Tuljapur, India.
UNWTO (2011) *Religious Tourism in Asia and the Pacific*. United Nations World Tourism Organisation (UNWTO), Madrid, Spain.
van Der Veer, P. (1988) *Gods on Earth: The Management of Religious Experience and Identity in a North Indian Pilgrimage Centre*. The Athlone Press, London.
Verdia, H.S. (1982) *Religion and Social Structure in a Sacred Town: Nathdwara*. Researchco Publications, Delhi, India.
Vidyarthi, L.P. (1961) *The Sacred Complex in Hindu Gaya*. Asia Publishing House, Bombay, India.
Vidyarthi, L.P., Jha, M. and Saraswati, B. (1979) *The Sacred Complex of Kashi: A Microcosm of Indian Civilisation*. Concept Publishers, New Delhi, India.
Wilbur Smith Associates (2005) *Action for Environmental Improvement of Alandi: Concept Plan*. Maharashtra Pollution Control Board, Alandi, India.

4 Human Sanctuaries Can Be Created Everywhere: Pilgrimage, Tourism, and Conservation in Vrindavan, India

Joshua Nash[1,2*]

[1]*Associate Professor, Aarhus Institute of Advanced Studies (AIAS), Aarhus University, Aarhus, Denmark;* [2]*Postdoctoral Research Fellow, University of New England, Armidale, Australia*

Introduction

The pilgrimage town of Vrindavan is located 150km south of Delhi, the national capital of India. A town of approximately 65,000 people, Vrindavan has more than 5500 temples dedicated to Krishna, the tenth incarnation of Vishnu, one of the trinity gods in the Hindu pantheon. The history of the town dates to the early 1500s when leading Vaishnava gurus who were followers of Vishnu and of the Bhakti (devotional worship) movement arrived on the banks of the Yamuna River in search of places related to Krishna's legends. The gurus discovered Vrindavan as a place with geographical features such as rivers, hills, lakes, and forests that characterized the setting of the pastimes of Krishna. They encouraged their disciples to visit these features and articulated the routes, itinerary, and format of pilgrimage in this landscape (Entwistle, 1987). Since then, Vrindavan has continuously evolved as a place of pilgrimage. Its expansion from a small rural settlement comprising of forest groves and meadows that attracted pilgrims to Vrindavan in the past into a place of extraordinary religious heritage of temples, havelis, pilgrim lodges, ashrams, and other religious infrastructure necessary for pilgrims is well-documented (Entwistle, 1987; Shinde, 2008).

Contemporary influences of pilgrimage, tourism and expanded development, however, have overwhelmed the town. In 2017, 13.78 million people visited Vrindavan (Tourism, 2018), almost double what was reported a decade before (Shinde, 2008). Most of the natural landscape has given way to housing and infrastructure developments. At the same time, the ancient temples in the older parts of the town have become increasingly dilapidated, the streets clogged with human and vehicle traffic, the general ambience of the town has been disturbed, and the environmental conditions have become very poor (Haberman, 2006; Shinde, 2008). Vrindavan has begun to take on complications reminiscent of urban environmental problems, and many authors have repeatedly noted how Vrindavan pilgrims have been disappointed that they are unable to view the imagined landscape depicted in religious scriptures (e.g. Haberman, 1994; Sullivan, 1998). And yet, 'the unchanging image of the place as Krishna's land and its transcendent qualities prevail over the physicality

*jnash7@une.edu.au

of environmental problems' (Shinde, 2011, p. 460). (Haberman, 1994) p. 48) notes that 'regardless of what has actually survived in this age of deforestation, [this urban landscape] is conceived of as a forest'. This imagination has continued to inspire environmental conservation efforts in Vrindavan. For instance, religious institutions and social movements in Vrindavan have focused on reforestation through the Vrindavan Conservation Project funded by World Wide Fund for Nature – India. In addition, this project, along with other non-governmental organizations (NGOs) such as Friends of Vrindavan, has focused on creating awareness of environmental issues which were first noted by religious gurus and devotees.

Moving beyond the ongoing discussions and lament about the degradation of the physical environment of the town, this chapter takes a philosophical perspective and focuses on the imagination of Vrindavan as a sacred place that inspires a distinct eco-spirituality. The imagination of the idyllic setting of mythical Vrindavan, as articulated by the Vaishnava poet-saints in their devotional literature provides deep insights into the relationships between humans, nature, and the divine. In creating the religious framework of theology and practice for Vaishnava devotees, the poet-saints eulogized Vrindavan as a forest of Krishna's pastimes and the terrestrial playground of the divine couple, Krishna, and his eternal consort, Radha. Just as the Hindu god Shiva is 'the destroyer' and the Hindu god Brahma is 'the creator', Krishna is seen as 'the enjoyer'. Vrindavan and its forest groves were created to serve as an earthly home for the divine couple to enjoy. And in the same way that Krishna and Radha took on human form to experience the bliss and the sanctuary of the gardens of Vrindavan, there is also the possibility for humans to access the same type of blissful experiences in this sanctuary.

The idea of Vrindavan as a sanctuary that nestles religion-spirituality-ecology connections is at the centre of this chapter. It argues that a true pilgrim/devotee of Krishna visiting Vrindavan and worshipping this sacred landscape can internalize these connections through meditative practices and spiritual alignment, evolving this landscape into a human sanctuary. The idea of Vrindavan as the 'human sanctuary' was coined by local Vrindavan ecologist Shri Sevak Sharan (Sevakji) who has been working towards environmental conservation in Vrindavan since the late 1970s. He elaborates on this through his Vrindavan Ecological Concept (VEC), which in turn is based on imaginations of Vrindavan in devotional literature. He argues for practical, spiritual-based action and contemplative practice (*sādhanā* and *chintan*) to realize this concept. As a westerner seeking the melding and meaning of religion and ecology I eventually took Sevakji to be my teacher. As his student over the last 20 years, I have imbibed some of the VEC concepts in my becoming a pilgrim of religion and ecology which I reflect on in this chapter. This reflection is part of my ethical responsibility towards my guru who desired that I 'become a real environmentalist for Vrindavan' and through my *sādhanā* (i.e. spiritual and meditative practice) that I share the eco-spiritual understandings from a sacred place like Vrindavan to a wider world. My view is that sanctuary exists both in the physical and metaphysical sense. Based on the eco-harmony found in and principles learned from Vrindavan, humans around the world should be able to create human sanctuaries anywhere.

In this chapter, I first refer to the Vaishnava devotional literature which provides some commentary on the environment of the divine place dedicated to Krishna and creates a certain imaginary of Vrindavan. A major focus is on a philosophical poetic work called *Hit Chaurāsī Pad* (*Eighty-four Stanzas*) written by Hari Vansh Goswami, a poet-saint from the 16th century. In this work, one can find the ecological, theoretical, and scriptural explorations of Vrindavan as a sanctuary place as well as a metaphysical concept. Then I turn my attention to Sevakji's articulation of the VEC through an eco-spirituality approach, which to me is an excellent vantage point to examine deeper issues beyond pollution control, disease, waste amelioration, and environmental management. While doing so, I explain the ways in which the human-nature-divine relationship is framed through Vaishnava religious practice in the worship of Krishna and Vrindavan. Following these explanations, I dwell deeply on how the teacher-student (*guru-shishya*) relationship that developed in the ashram of Sevakji has helped me to understand the deep ecology of Vrindavan and internalise the concept of human sanctuary by focusing on the ideal Vrindavan of the mind, spirit, and devotion. Finally, I return to the interpretation of this philosophy for environmental conservation in physical Vrindavan, and conclude the chapter by considering how ecological thinking and environmental action have become integral to how the VEC, Sevakji's philosophy, and the human sanctuary idea can potentially be manifested and extrapolated through the actions and behaviour of religious communities, devotees, and pilgrims in the city of Vrindavan.

The Environmentalism in Vaishnava Devotional Literature

The 16th-century poet-saints through their scriptures created an image of Vrindavan as a place of divinity. A seminal work in this regard is the *Hit Chaurāsī Pad* of a poet-saint Hari Vansh Goswami (1502–1552). Like many saints of the time who were inspired by Vaishnava devotional worship, which encouraged visiting the newly discovered land of Krishna's pastimes, Hari Vansh Goswami was also attracted to the divine forest groves of Vrindavan. As a devotee of Krishna, he realised the potential fruits of human life and scribed detailed synopsises of his realizations of the pilgrimage landscape that Vrindavan was. Later, he founded a new tradition of worship in Vrindavan through his Radha Vallabh temple (see Snell, 1984, 1991 for more details of Hari Vansh's life). The major position taken by Hari Vansh Goswami in his poetry is that Vrindavan is the terrestrial stage where the male Krishna and the female Radha, in their human forms, represent two equal and opposite forces of life energy, consciousness, or *brahma* (spirit). The environmental imagery of verdant forest groves of Vrindavan provides the setting for the celestial pastimes of the divine couple Radha and Krishna. Krishna and Radha are worshipped for sporting amusement and aesthetic experiments into the science of *ras* (pleasure). A few verses from *Hit Chaurāsī Pad* translate as:

> 'Come, wise Radhika! For your sake Shyam has arranged a round dance, a store of joy, on the bank of the Yamuna
>
> 'In that most pleasing place near the vamshivata a soft breeze blows from the [sandal-clad] Malaya mountain, yielding all joys
>
> 'the forest is strongly fragrant with half-blown jasmine, and there is bright moonlight in the full-moon autumn night
>
> 'lady! Experience this ocean of delight, rejoice with your arms joined around his neck. For Shyama's sport in the fresh bower is worthy of the world's praise!'

While the physical forms of these deities are engaged in the act of pure enjoyment and for the achievement of bliss, they also represent a cosmic drama, where Krishna and Radha make each other complete. Without one, the other is incomplete. Similar duality exists between deity and devotee, which duality creates the divine experience within the human-nature relationship. By being aware of this duality, humans can understand the relationship between life and nature. For example, in one of the forests, one tree has white and dark branches intertwining with each other; leading devotees believe that this represents the embrace of Krishna and Radha (Shinde, 2008).

Vaishnava poet-saints not only portrayed the divinity of landscape called Vrindavan but also commented on its transformations into a physical place of worship. For instance, another poet-saint of the 17th century, Hariram Vyas, criticized the construction of temples in forested areas:

> The Kaliyuga has really arrived ...
>
> Why have you left me alive to see this day when trees are being cut [in Vrindavan] and the sacred land is being dug for making way for temples [in Mathura]?
>
> (Translated from Hindi) (Shinde, 2008).

Another poet-saint, Biharindas lamented:

> These people are stupid – leaving the importance of Braj-raj (soil of Braj), they are making temples of stones – they are doing all wrong things.
>
> Due to their greed they are not able to gain their guru's best wishes and therefore they have become sick with greed,
>
> Ministers have become bad advisors; they have cut the bowers and brought in stone thereby making the kunj [forest grove] non-productive.
>
> They have cut the trees and loaded logs on bullock carts and the axe is seen. More so by driving the bullock carts in the tender soil of Braj, they have made it like fine flour.
>
> (Translated from Hindi) (Shinde, 2008).

Fig. 4.1. Sevakji with the author, Vrindavan, December 2003 (Photo courtesy of the author).

Sevak Sharan as an Eco-guru and the VEC

Much of the existent devotional literature, particularly that of Hari Vansh, demonstrates the importance of the lived, natural experience, the relationship between nature and culture, and how poetry, art, and the ancient perspective and role of the natural and religious ritual depict the dialectical interplay of the human being with nature. It is the lineage of Hari Vansh Goswami and the Radha-Vallabh Sampradaya (temple tradition) that he founded with which Sevakji is associated (Nash, 2015). Following on the philosophical tenets of his ancestral gurus, Sevakji reinterpreted the environmental situation in Vrindavan and formulated the VEC, which I describe below.

My association with Sevakji began in the late 1990s in Vrindavan during a brief stint as a student environmentalist and aspiring devotee (see Nash, 1998). I also made six visits to Vrindavan between 2003 (see Fig. 4.1) and 2008, during which time I worked intently with Sevakji to synthesize several disparate strands of spiritual practice (*sādhanā*) and the VEC.

Sevakji grew up in Vrindavan, and having been steeped in the knowledge and experience of his parents' generation of stringent and orthodox temple worship and spiritual practice, he had deep questions regarding the relevance of scripture, practice, and devotion to the practical aspects of environmental conservation. He was familiar with the ecological vision as found in Hindu religion which considered nature and the environment as divine and sacred (Chapple and Tucker, 2000; Dwivedi, 2000). Having read devotional scriptures dedicated to Krishna and Vrindavan, he became aware that the worship of Krishna and Radha was filled with ideas about conserving the environment. And yet he daily witnessed the degradation of the natural environment. Sevakji was trying to find answers as to why this place – historically a highly charged place of pilgrimage and ritual – had become so neglected ecologically, culturally, and religiously. This damage had occurred despite there being ample money, dedication, and local and international interest to remediate the squalid environmental state of the town. This outcome initially led to the instigation of several cultural and ecological conservation programmes, including preserving the sacred bathing *ghats* by the Yamuna River in the early 1980s, education programmes aimed at encouraging reforestation, and ultimately the founding of the World Wide Fund for Nature – India's Vrindavan Conservation Project (Nash, 1998; see also Kumar, 2010, Kumar, 2019; Sharma, 2010; Kumar and Jacobsen, 2017).

In parallel with the outward movement of establishing the environmental and social movement of Vrindavan conservation, Sevakji set up what was to become the prototype for his vision on the then remote pilgrimage path that surrounded the town. He established a nursery in 1980 on land he inherited from his family. Slowly, this single-acre hermitage where Sevakji still lives became the training ground for disseminating knowledge and experience about the principles of nature and spirituality that have evolved from years of practice and ecological work in Vrindavan. Integrating philosophical tenets from scriptures into his practical work, Sevakji proposed the VEC model that would help in addressing some of the environmental issues in Vrindavan.

VEC: Seven Principles of Human Ecology and Spirituality

The VEC comprises of seven principles of human ecology that Sevakji derived from his study and rootedness in Vrindavan. The first principle is spirituality – the realization that humans exist in a world brought about by something greater than themselves. The world and everything in it are divine. The second principle is that nature is the outer manifestation of this spiritual inner truth. All creatures learn from nature, and human beings need to realize that they give and take from nature. The third principle is that human-nature interactions are what create human culture. These cultures in turn create tangible and intangible heritage – the fourth principle, through which specific cultural expressions are maintained over time and are deeply embedded in people's dealings with the environment through music, art, architecture, food, and lifestyle. Pilgrimage and experiencing sacred places, the fifth principle, are methods through which one can understand differences and similarities between cultures across time and environment. These mobilities help people realize that human beings ultimately share the same strivings, problems, needs, desires, and will to find peace in life. From this realization comes concerns over human welfare, the sixth principle, and associated social and environmental ideas, programmes, and activities. When individual humans are in harmony with themselves, they can experience similar feelings with their close social group, greater society, and nature. They will then also understand how these positive emotions and interactions can be commonly shared universally all over the world. All such interactions expand to the ecology of the world – the seventh and overarching principle. Thus, if we follow the seven principles in succession, this world represents an extension of human ecology which begins with a spiritual understanding of oneself. In this way, Vrindavan is a microcosmic representation of the world.

Underpinning these seven principles of human ecology is the notion of consciousness. The VEC can be realized through developing five types of consciousness that follow on from Hari Vansh Goswami ecological-cum-scriptural injunctions and are evident from the life and pastimes of Radha and Krishna. The first consciousness is Krishna Consciousness. Many religious gurus interpret Krishna legends as socially and ecologically focused pastimes and thus portray him as an ecologist and as an upholder of dharma. For instance, Shinde, 2011, p. 458) notes how a Vaishnava guru in Vrindavan highlighted the need to contextualize environmental problems within Krishna worship by suggesting that 'the presence of the serpent, Kaliya, was like pollution in the Yamuna River. By killing Kaliya, Krishna actually destroyed the pollutant in Yamuna and not merely the physical form of Kaliya'. The second consciousness is Radha Consciousness, where one can learn practically through the possibility of grouping with others intimately. Radha and Krishna demonstrated and manifested divine love as known from several legends where they engaged in circles of dance on the banks of the Yamuna River (Entwistle, 1987). In the third consciousness, Yugal (conjugal) Consciousness, people perceive an awareness of the relationship of divine opposites regarding nature, love, relationships, everything in the cosmos. This awareness is enabled through the fourth consciousness, Vraj (or Braj) Consciousness – Krishna's interaction with the people and environment of Vraj, the geographical location where Krishna manifested his childhood and adolescent pastimes. This interaction demonstrates the possibility of how to interact with the generally unbalanced outer society while maintaining one's inner balance through practical ecological and socially balanced behaviours (Entwistle, 1987). Finally, Nature-Vrindavan Consciousness expresses ways of living in harmony with nature. Specifically, the beautiful environment of Vrindavan is the place in which Radha and Krishna undertook their divine activities and affairs.

According to Sevakji, in Vrindavan one can find an integral and pervasive connection between humans and the divine, and hence Vrindavan has the potential to represent in a microcosm the prospect that the entire world could in principle become a human sanctuary.

Following this premise, I conceive of 'human sanctuary' as a place-state where humans can live in a vibrant, non-static, and peaceful consonance with themselves as individuals and in an intimate and loving partnership with the greater human society and with nature. During my stay at Sevakji's ashram in Vrindavan, I was able to live and experience these eco-spiritual lessons more closely, as I illustrate below.

Nature Is-as Teacher: Living and Imbibing Eco-spiritual Lessons From an Eco-guru

I have visited Vrindavan many times in the last 20 years and stayed at Sevakji's ashram, affectionately known as Latā Bhavan or the Abode of Fauna on the *Parikramā Mārg* in Vrindavan (Fig. 4.1). Over this time, I engaged in developing the *guru-shiṣhya* (teacher—student) union in both my working life as an academic and in my perpetual striving for fulfilling the path that Sevakji had shown me. I could see that Sevakji was not merely an ecologist who possessed some clear ideas about how the Vrindavan environment could be remediated. Rather, he had synthesized a theory of the relationship between the human, nature, society, and the Divine which had yet to be written or compiled in any significant way. During my experiences with Sevakji, what I discovered and was taught was a vast amount of philosophical, empirically scientific, as well as personal *sādhanā*-based research that was waiting to be written, put into practice, and shared with the world, particularly about the VEC as a philosophy building on the human sanctuary idea.

What became apparent to me was that the more I looked at the links between the elements in Sevakji's system of thought, the more difficult it became to synthesize his system. The more intense our *guru-shiṣhya* relationship became, the more the realizations grew about how this personal Vrindavan environmental awareness creation, *prakriti-chintan* (meditation on nature), *prāṇa-sādhanā* (meditation on life energy), *kāl-sādhanā* (meditation on time and its movement), and spiritual schooling interaction were achievable for a seeker. A tangible opportunity to articulate this thinking came when Sevakji urged me to write and publish a synthesis of the research he had conducted in developing VEC. Initially he asked me to compile a small booklet that introduced some elements of the VEC so he could use it as a simple teaching tool for the many visitors and devotees who came to visit him in his ashram.He believed that this booklet would satisfy the intellectual side of any seeker.

Sevakji continually shared with me that his thought processes encompassed four major philosophies or, as he called them, *driṣhṭi*, or visions. The first was the ancient or Vedic vision of India that most Indians would have known from scriptures of Vedas that talk at length about interactions between humans, deities, and nature. The second was the Rādhā–Krishna vision, which is essentially embedded in the worship of Krishna and Vrindavan and what I have come to term 'the self-vision'. The third was the Gandhian vision, attributed to the simple living advocated by Mahatma Gandhi, and what I term the local or sustainable vision. The fourth was the modern scientific vision that is embraced by contemporary ecologists and environmentalists, which, according to Sevakji, are like modern-day yogis (spiritual practitioners) who keep close and continue the cycle from the modern scientific vision back to the ancient Vedic and natural vision. While such thinking is inspired by immersion into the physical and cultural setting of Vrindavan, these concepts are universal and could be taken by pilgrims and seekers back to the places from which they came. That is, Sevakji believed that these principles, when written in booklet form, would provide lessons from Vrindavan that could be extrapolated and practiced elsewhere.

Another milestone in my learning from Sevakji came in 2004 when he said something to me that changed the direction of our relationship and allowed the next step in my practice to proceed: 'You should become serious about *sādhanā*.' I had apparently become a little careless in my awareness practice, and with this statement realized I had not been appreciating the gravity of the *sādhanā* process in which I had been engaging and the gravitas expected by Sevakji with which to apply myself. I was resident in the *gurukula* (the house of guru/teacher). Because of my exposure to Gauḍīya Vaishnava ideas, Bengali Krishna worship associated with the medieval saint Chaitanya, and its relationship to *guru-tattva*, or the truth or essence of the teachings of the teacher, I began to appreciate the rareness of having the privilege to 'live in' with such a teacher and to be carefully watched and trained by this guru. The instruction I was receiving was not only based on being exposed continually to his intellectual teachings – the focus was intently on the significance of *nature* as our principal teacher.

In certain takes on Krishna philosophy, there is the book *bhāgavat* and the person *bhāgavat* (*bhāgavat* meaning devotee, worshipper or follower of Bhagavanta or God). The book *bhāgavat* is believed to be a scriptural manifestation of the Divine and its reading prepares the reader to have experiences with the Divine, while the person *bhāgavat* refers to holy people who can lead a person to the Divine through their teaching, spiritual guidance, and counsel. Through both book *bhāgavat* and person *bhāgavat* people can arrive at the mystical *bhāgavat*, or the Divine. The goal or intent encouraged by many seers is to listen to and think about the *bhāgavat* and Krishna's life and teachings 24/7 from the lips of a genuine guru, a person *bhāgavat*, or a devotee who has realized that Krishna is present in Vrindavan. It is believed that guru is the physical form of God and the subtlest form of guru is God.

For me, both Sevakji and his ashram in Vrindavan were a guru. Sevakji's ashram had itself become a micro *dhām*, holy place, and a *tirtha-sthān* or a confirmed place of pilgrimage, because of the individuals, yogis, and sages who resided there over the past century or more and added layers of divine practices and interactions between humans and nature in creating the sacredness of the place. I believe that the *dhām* (place), *prakriti* (nature), and environs that Sevakji had set up in his ashram provide a small-scale case study-cum-template of larger scale human—human, human—nature, and human—divine interactions that one can find within the greater cosmos. I learned that through self-introspection and *chintan* or contemplative meditation, which creates awareness of one's self and own life energy or *prāṇa*, one could see, experience, and learn from a guru in nature and become cognizant of one's self in world. In this system of thought, nature is the curtain of the Divine and an inherent and intrinsic guru in the cosmos and the biosphere. Vrindavan as a place of learning *is* guru. The human sanctuary is not only scriptural. Interactions between two human beings represented as teacher—student interaction can help people understand and live individual and personal human—nature interactions in a more balanced way. Thus, people are to live in harmony with nature and others and come to this self-realization in order to experience the Divine. Because Vrindavan is the model of human—human and human—nature interactions, extending the ways in which the people/government in Vrindavan live their lives and applying them to other contexts the world can eventually become Vrindavan and experience the Divine. This is core to the thesis that human sanctuaries can be created everywhere.

As an aspiring *sādhak* or devotee, I wondered: what did this realization mean to me, and how could I put this into practice? How can one listen to book *bhāgavat* (sacred scriptures) 24/7? Does a person always have to be around the person *bhāgavat* (saint)? What is the relationship between *vāṇī* (the message of the teacher) and *vapu* (the form of the teacher) of the person *bhāgavat*? And where can one find a guru in this material world anyway? It became clear to me that nature is guru, the original guru whereas the place is a guru when one is in that place such as Vrindavan. So, if we listen to nature speaking, which we *can* do 24/7, then we *are* hearing the Divine all the time. One does not actually need to have a human guru, nor do I believe one needs a guru connected to any manifestation of the Krishna religion. Nature is teacher; Radha and Krishna are both in the world and accessible, as well as being distant, transcendent, and unattainable; and people can live a life in bliss with ourselves, our partner and intimate contacts, nature, and society. That is, we can simultaneously live in all types of consciousness: Krishna (self) consciousness, Rādhā (partner-other-intimate) consciousness, Vrindavan (nature) consciousness, and Vraj (societal) consciousness.

Because of this change of focus from the human to the natural, a guru could be experienced anytime and anywhere. For the several years I spent in the ashram of my teacher, guru, friend, and co-researcher, Sevakji was my form – *vapu* – and his message – *vāṇī* – to me was simple: be a real environmentalist for Vrindavan. While he did not tell me how to be or become such a devotee – that is part of my own lifework – what is clear over an almost 20-year connection to Vrindavan is that both the physical place called Vrindavan and the 'spiritual Vrindavan' are important. Or, as I have written elsewhere in opposition to flippant positions about the role of the environment in Vrindavan, 'Idealized stances including the idea that only the transcendental Vrindavan matters as opposed to managing and addressing the current ecological state of the terrestrial Vrindavan are not taken seriously' (Nash, 2012, p. 109). Vrindavan is Rādhā and Krishna's *Rāsa-līlā* or cosmic playground. It is meant to be a terrestrial representation of something divine. However, it is still important to remember that in its physical and material form, it is a polluted and consumptive religious dreamland that is rapidly urbanizing.

As a writer, one element of my 'being a real environmentalist for Vrindavan' means committing to paper such topics. This chronicling is essential to document specifically not only the history of my interaction with Sevakji and the Vrindavan environmental movement, with which I am still involved, but also how personal takes on knowledge transfer and *sādhanā* practice can be expressed more generally. Despite modern technological advancements and their application in devotional circles like mass television preaching, I contend that the role of guru as teacher and the importance of their physical presence when imparting to disciples the possibility of attaining developed self-knowledge has not fundamentally changed. Because the Krishna sect's intricate system of knowledge – *Veda* –is traditionally represented in religious institutions and systematised patterns of belief, the role of the physical guru in the dissemination of knowledge becomes vital. Gurus habitually live in remote and simple ashrams where seekers come to visit and seek out knowledge and enlightenment. It is in these ashrams and the philosophies presented by gurus where a clearer appreciation of the relation between religion-philosophy-scripture and nature can be arrived at. I claim this is what Krishna came to teach as guru, namely that the whole cultural and natural world, when seen with a divine perspective, *is* Vrindavan. That Vrindavan is a human sanctuary and that Vrindavan is and can be anywhere.

Exiting the Sanctuary, Re-entering the World

From its rural beginnings, Vrindavan has evolved into a popular Hindu pilgrimage town and its environmental history is well-documented (e.g. Entwistle, 1987; Shinde, 2008; Kumar, 2019). In the last four decades in particular, the town has suffered intense environmental degradation due to a phenomenal increase in numbers of pilgrims. The practical groundwork and governmental lobbying of, for example, the World Wide Fund for Nature – India has not been able to produce the intended results such as reforestation and broader societal awareness of environmental issues. Although small-scale projects and initiatives like Friends of Vrindavan have brought about temporary outcomes at the micro-level (such as cleaner streets), larger governmental projects, such as the many proposals to remediate the Yamuna River, have not lived up to their requisite potential on a macro-level.

Considering the increasing religious tourism and its impacts, it is not clear whether there could be positive social and ecological change to the sacred landscape of Vrindavan. How would worship of Krishna as propagated by different Krishna sects and denominations lead the interactions of devotees with the sacred environment? I have elsewhere argued that 'according to scriptures from all of the prevalent temple traditions, Vrindavan should render the epitome of environmental awareness and thus divine awareness real' (Nash, 2015, p. 60). However, this awareness is not currently visible in this temple town. What is the role of the media and the internet in mustering support and doctrinal change in terms of religiosity and environmental change? These are large and perplexing questions, the answers to which do not only belong in the written or even thought and emotional realm, but in the realm of focused action.

However, this chapter has moved beyond the physical and material world of Vrindavan and presented a clear hypothetical statement on the idea of human sanctuary. This idea is rooted in religious thought and *chintan*, or contemplative practice, as well as on documenting relevant social and theological history and practice associated with the VEC and Vrindavan environmentalism. In principle, the conceptualization of human sanctuary within the framework of the VEC is inspired by the human—nature—divine—relationships that can be experienced in a sacred place like Vrindavan. Vrindavan is both a real place and a concept. The town's physical environment and devotional culture is important in developing 'a correct vision to make the Vrindavan = Nature + Divinity scenario a reality' (Nash, 2015, p. 60). I argue that by synthesizing a nature-focused, *dham-sadhana*-centric process a pilgrim and a seeker can achieve a divine vision and when such a person 'looks to nature with divine vision, that place becomes Vrindavan. And that place, which becomes Vrindavan, the place where there is consonance in the Humanity—Nature—Divinity trinity, becomes a *Human Sanctuary*, a place of sanctity, solace, and the sacred' (Nash, 2015, pp. 60–61). It is the how, where, why, and when which seem to be the major challenges of the next stage of Vrindavan conservation.

Using the template of Vrindavan, I have put forward ways where practical and personal lessons learned (e.g. personal spiritual practice in line with ecological ideals) are generalizable and can be applied beyond this northern India example and to religion and tourism locations elsewhere. Similar stories of devotion exist in other sacred landscapes such as in Bodhgaya, India, as applied to Buddhism (e.g. Geary, 2008), and to other gods such as Shiva in Varanasi, India (e.g. Singh, 1997). The appreciation of *nature as teacher* and *nature as a curtain of the Divine* is essential for understanding Sevakji's teachings. Vrindavan is not merely a poetic metaphor or a metaphysical reality to be experienced in some transcendental realm beyond this mortal, material existence; it is an actual physical historical place, where Radha and Krishna, after carrying out their pastimes more than 5,000 years ago, still live them out through their devotees. By addressing the human sanctuary idea, it is hoped this most venerated and holy place will move towards a sober preservation of its ecological and religious integrity.

References

Chapple, C.K. and Tucker, M.E. (eds) (2000) *Hinduism and Ecology: The Intersection of Earth, Sky and Water*. Harvard University Press for the Center for the Study of World Religions, Cambridge, Massachsetts.

Dwivedi, O.P. (2000) Dharmic ecology. In: Chapple, C.K. and Tucker, M.E. (eds) *Hinduism and Ecology: The intersection of Earth, Sky and Water*. Harvard University Press for the Center for the Study of World Religions, Cambridge. Massachsetts, pp. 3–20.

Entwistle, A. (1987) *Braj: Centre of Krishna Pilgrimage*. Egbert Forsten, Groningen, The Netherlands.

Geary, D. (2008) Destination enlightenment: Branding Buddhism and spiritual tourism in Bodhgaya, Bihar. *Anthropology Today* 24(3), 11–14. DOI: 10.1111/j.1467-8322.2008.00584.x.

Haberman, D.L. (1994) *Journey through the twelve forests: An encounter with Krishna*. Oxford University Press, New York.

Haberman, D.L. (2006) *River of love in an age of pollution: The Yamuna river of Northern India*. University of California Press, Berkeley/Los Angeles.

Kumar, S.S. (2010) *Bhakti-the Yoga of Love: Trans-Rational Approaches to Peace Studies*. LIT Verlag, Münster.

Kumar, S.S. (2019) *Vrindavan's Encounter with Modernity: Changing Environment and Life-worlds in an Indian Temple Town*. LIT Verlag, Zürich.

Kumar, S.S. and Jacobsen, E. (2017) Heritage and environment: visions of past and future in the Indian temple town Vrindavan. *Peaceworks* 7(1), 115–130.

Nash, J. (1998) *Vrindavan Conservation – A Perspective*. WWF India, New Delhi, India.

Nash, J. (2012) Re-examining ecological approaches to Vrindavan pilgrimage. In: Manderson, L., Smith, W. and Tomlinson, M. (eds) *Flows of Faith: Religious Reach and Community in Asia and the Pacific*. Springer, Dordrecht, The Netherlands, pp. 105–121.

Nash, J. (2015) Vrindavan: the human sanctuary. *Journal of Vaishnava Studies* 24(1), 55–66.

Sharma, M. (2010) The Vrindavan conservation project. *Economic and Political Weekly* 45(36), 59–66.

Shinde, K.A. (2008) The environment of pilgrimage in the sacred site of Vrindavan, India. PhD. Monash University, Melbourne, Australia.

Shinde, K.A. (2011) What are charitable trusts doing in religious tourism? Insights from an Indian pilgrimage site. *Tourism Planning & Development* 8(1), 21–36.

Singh, R.P.B. (1997) Sacred space and pilgrimage in Hindu society: the case of Varanasi. In: Stoddard, R.H. and Morinis, A. (eds) *Sacred Places, Sacred Spaces. The Geography of Pilgrimages. Geoscience & Man*. 34. Geoscience Publications, Baton Rouge, Louisiana, pp. 191–208.

Snell, R. (1984) The Caurasi Pada: a Radhavallabhi text in medieval Hindi. Doctoral dissertation. SOAS University of London.

Snell, R. (1991) *The Eighty-Four Hymns of Hita Harivamsa: An edition of the Caurasi Pada*. Motilal Banarsidass, New Delhi, India.

Sullivan, B. (1998) Theology and ecology at the birthplace of Krsna. In: Nelson, L.E. (ed.) *Purifying the Earthly Body of God: Religion and Ecology in Hindu India*. SUNY Press, New York, pp. 247–268.

Tourism, U.P. (2018) Annual tourist visits statistics for Uttar Pradesh, 2017. Available at: http://www.up-tourism.gov.in/pages/top/about-up-tourism/year-wise-tourist-statistics (accessed on 11 October 2019).

5 The *Beyul*: Sherpa Perspectives on Landscapes Characteristics and Tourism Development in Khumbu (Everest), Nepal

Sanjay K. Nepal[1]*, Yang Mu[1] and Po-Hsin Lai[2]
[1]University of Waterloo, Waterloo, Ontario, Canada; [2]University of Newcastle, Newcastle Business School, Australia

Introduction

Studies on landscapes have been of interest to scholars in a variety of disciplines, including geology, geography, planning, ecology, environmental studies, history, architecture, archaeology, and anthropology (Tilley and Cameron-Daum, 2017). Landscapes are dynamic and subject to changes in contemporary environmental, economic, societal, and technological developments and societal values and culture that give different meanings and importance to its multifaceted domains (Terkenli, 2004). Within geography, landscapes are defined conceptually as part of nature, environment, region, place, and space (Setten, 2006). A growing body of literature within geography is focused on sacred landscapes, which are found in mainstream faiths, indigenous beliefs, and non-religious contexts (e.g. Lane, 2002; Bremer, 2006; Perriam, 2015; Jazeel, 2018). These landscapes are often protected as sanctified sites because they are viewed as places of the living expression of certain gods, deities, or spirits worshiped or honoured by local residents (Dudley *et al.*, 2009). Although different cultures interpret the word 'sacred' differently, sacred landscapes around the world are often associated with strict behavioural restrictions, a sense of separateness, and strong emotion-oriented and place-bound characteristics (Hubert, 1997; Levi and Kocher, 2013). They are sites reflecting humanity's various forms of interaction with supernatural entities (Stump, 2008). Sacred landscapes, therefore, are objective and subjective representations of embodied practices and experiences observed in natural—cultural environments. These practices and experiences are increasingly coming under the influence of tourism (Knudsen *et al.*, 2008).

The purpose of this chapter is to examine local residents' perspectives of Nepal's Khumbu (Mt Everest) region and the role of tourism in shaping residents' perspectives of this sacred landscape. This region is believed to be a sacred *beyul*, or a hidden valley that acts as a refuge, by local Buddhist Sherpas, and has undergone tremendous environmental, economic, and social changes due to the increased popularity of trekking tourism in the area. This chapter is based on data and information collected during two visits to the study location in 2014. Preliminary observation of the geographical and human contexts, main settlements, and the dynamics of the trekking tourism industry were made during

*Corresponding author: snepal@uwaterloo.ca

the first visit in April and May. During the second visit (October and November), 33 in-depth semi-structured interviews were conducted in 11 villages (Phakding, Monjo, Namche, Thame, Thamo, Khumjung, Khunde, Tengboche, Deboche, Phortse, and Pangboche). The interviewees were selected using a purposive snowball sampling technique. All 33 participants were of Sherpa ethnicity between 19 and 78 years old; more than half (20) were between the ages of 31 and 50. Interviewees included ten 'religious informants' (eight male, two female), either currently serving or who had previously served in monasteries and nunneries, 18 tourism entrepreneurs (14 male, 4 female), and five farmers (two male, three female). Qualitative content analysis and coding techniques were used in analysing the interview scripts. Respondents' names presented in the findings below have been changed to protect their identities.

Sacred Landscapes

Existing literature on landscape studies reveals two main approaches used by researchers in interpreting a landscape: 1) a morphological approach which emphasizes the tradition of quantification and classification of landscape (Mücher *et al.*, 2010), and 2) a humanistic approach which focuses on socially constructed elements in landscape (Knudsen *et al.*, 2008). The humanistic approach is applied in a wider range of work and is place- and context-specific. Landscape is viewed from this approach as more than just a physical form, and is considered as a complex expression of a dynamic system which involves the area's culture, custom, social structure, and values (Knudsen *et al.*, 2008). This humanistic approach also incorporates the 'spiritual views' in interpreting people—environment relations (Stokols, 1990), where the landscape is viewed as providing important contexts for cultivating human values and spirits.

Some scholars have suggested that the study of tourism cannot be separated from the study of the landscape, since the latter creates the geographical, historical, and cultural context for tourism (Knudsen *et al.*, 2008). Distinct features, such as the shape of a mountain that appears to resemble a human form, endow tourism landscapes with symbolic meanings (Cosgrove, 1998; Knudsen and Greer, 2008). But cultural landscapes are also subsequently influenced by tourism (O'Hare, 1997), as local communities express and share their beliefs with tourists who are encouraged to view these landscapes as the locals do. Tourists may also influence local perspectives as they share their ideas of how they view and relate to these landscapes. Tourism landscapes are therefore arguably a sub-category of cultural landscapes and overlap and coexist with other categories of cultural landscape such as heritage and sacred landscapes (O'Hare, 1997).

Although different cultures have different understandings of what constitutes the 'sacred', there seems to be a universal understanding of the concept of sacredness. The English word 'sacredness' is derived from Latin (*sacer*) and is generally defined as 'restriction through pertaining to the gods' (Hubert, 1997). This definition suggests that the concept of 'sacred' implies strict prohibitions and restrictions on behaviours. In other words, something is sacred because certain sets of rule are observed in relation to it. Hubert (1997) stated that concepts of separateness (e.g. sacred vs profane), respect, worship, and rules of behaviours are common to sacred sites in different cultures. The restrictions on human behaviours at sacred sites help define the relationship between humans and deities or powers. It is usually believed that superhuman powers can be protective and assisting, but they can punish and destroy at the same time (Horster, 2010). Therefore, certain rules are made to prevent offending the gods, who must be awed and appeased. These prohibitions of certain behaviours can be closely associated with religious activities and can also be tied to ritualized practices in daily life. In Christian belief, sacredness is very specialized and tends to have little connection with family relationships, mundane life, or with economic activities (Hubert, 1997).

But in many other religions, especially those with animistic characteristics, trees, springs, mountains, even an unmarked piece of rock can be sacred (Ormsby, 2013). This is particularly the case in the Khumbu region, where centuries old Tibetan Buddhism practices have affirmed the importance of nature and natural elements in manifestations of sacred forms and functions relevant to human

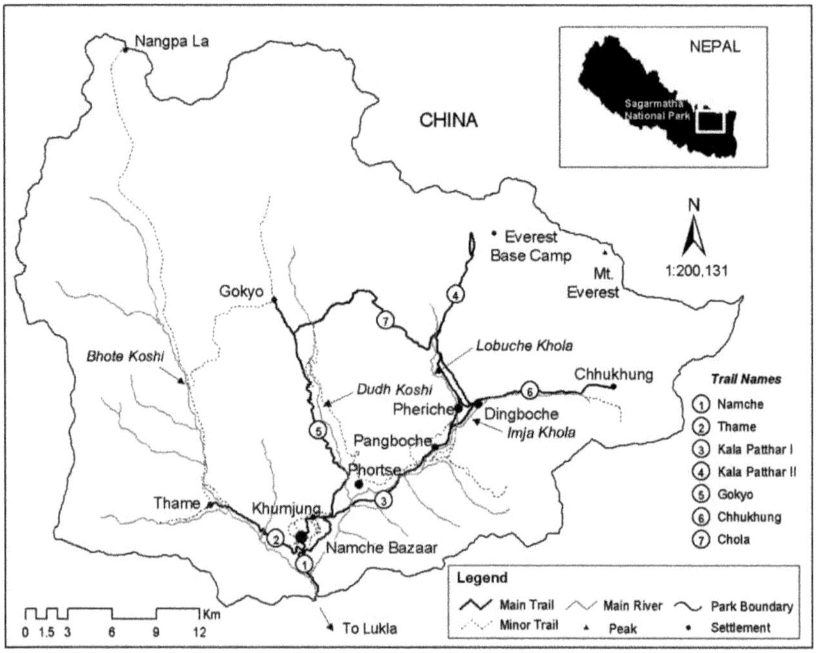

Fig. 5.1. A map of the SNP in Nepal showing trekking trails and main villages, and Mt Everest. Villages within park boundaries are part of upper Khumbu; villages south of the park boundary at Jorsalle (southern most tip of the trail on the map) are part of lower Khumbu (Nepal, 2003a).

emotions, experiences, spiritual values, and beliefs. Tourism has reinforced high sacred values placed on some aspects of the landscapes, such as monastic forests, but has also undermined or altered values (e.g. economic) of the same landscapes. Over time, sacred landscapes may be seen as important primarily for their ability to draw tourists and develop tourism and less so for their spiritual and cultural significance to local traditions and practices.

Khumbu and Its Sherpa Residents

The Khumbu region is located in north-eastern Nepal and commonly referred to internationally as the Everest region. The boundaries of this region share the outer perimeter of Sagarmatha National Park (SNP) and extend further south into its buffer zone. Khumbu is divided into the upper Khumbu region (areas within the park boundaries) and the lower Khumbu region (areas south of the park entrance at Jorsalle; Fig. 5.1). The mountainous terrain features eight peaks over 7000 metres, including Lhotse, Cho Oyu, Makalu, and the famed Mount Everest (referred to as *Chomolungma* by Sherpas and Tibetans and *Sagarmatha* by other Nepalese). The Khumbu region is rich in biodiversity, and hosts rare species of animals, such as *Ailurus fulgens* (red panda), *Panthera uncia* (snow leopard), and *Moschus leucogaster* (Himalayan musk deer). Due to its unique culture, landscape, and flora and fauna, the SNP was established in 1976 and was declared a UNESCO World Heritage Site in 1979 (Department of National Parks and Wildlife Conservation, 2017). For the same reasons, the region has also become one of the most popular tourist destinations in Nepal – in 2016, around 36,700 foreign trekkers visited the park (Ministry of Culture, Tourism, and Civil Aviation (MOTCA), 2017).

Approximately 3500 people currently live in the Khumbu region. About 90% of the population is Sherpa, one of many highland ethnic minorities in Nepal (Spoon and Sherpa, 2008).

Sherpa, or 'people of the east', are believed to have migrated from Kham in eastern Tibet to Khumbu more than 500 years ago. According to the religious leaders of Tibetan Buddhism, the Kham Sherpa migrated to Khumbu believing it to be a *beyul* created by the Buddhist saint Guru Rinpoche (also known as Padmasambhava), where Sherpa could use it as a sanctuary in times of wars and conflicts (Zangbu and Klatzel, 2011). By the mid-1500s, the Sherpa dominated most of the Khumbu region, and gradually built six main villages throughout the rugged landscape: Khumjung, Khunde, Namche, Thame, Pangboche and Phortse (Ortner, 1989). Several smaller villages, located higher than 4000 meters, were primarily summer herding villages that have now transformed into permanent tourist villages, such as Dingboche, Gokyo and Lobuche (Nepal, 2005). For centuries, Sherpa sustained their livelihoods by engaging in yak herding, farming, and trading with people in Tibet and with lowlanders in southern Nepal. The growth in trekking and mountaineering in the region in the last four decades has changed Khumbu's subsistence economy into a tourism-centric economy (Nepal, 2016).

Tourism Development in Khumbu Region

More than four decades of intensive tourism development in the SNP have led to various positive impacts in the region, such as development of modern infrastructure, increasing household incomes and improved living conditions, international exposure and recognition of local culture, and cross-cultural exchanges (Adams, 1992; Nepal *et al.*, 2002; Nepal, 2003b; Mu *et al.*, 2019). However, studies on tourism impacts in this region have tended to focus on the negative influences of tourism on the environment, particularly highlighting ecological problems associated with waste disposal, deforestation, and changes in land use and agricultural practices (Nepal, 2003b; Nepal and Nepal, 2004; Byers, 2005; Nepal, 2016). In recent years, research on the social and cultural influences of tourism has found that tourism has led to changes in Sherpas' lifestyle, economic patterns, education, marriage choices, and location of residence (Sherpa, 2008; Klatzel, 2009), with traditional language, family ties, and cultural norms and values being not as important to younger generations of Sherpas (Luger, 2000). Some scholars, however, have argued that the negative impacts of tourism on the local environment, culture, and society in the Khumbu region have been exaggerated (Stevens, 1993; Spoon and Sherpa, 2008), as the Sherpa people have demonstrated a strong resilience toward changes brought by outside cultural forces. The traditional relations of production, cultural order, and religious routines largely persist, although some of these might have been transformed through adaptations to modern technologies and ideologies. Similarly, Sherpas have been very successful in negotiating between tradition and modernity and have largely kept the essence of their culture intact despite their deep involvement in tourism (Adams, 1992; Nepal, 2015).

Sherpa Concepts of the *Beyul*

Most Khumbu Sherpa practise Nyingmapa, or 'the old translation school', of Tibetan Buddhism, while some Sherpa clans in other regions of Nepal follow the newer Gelug School of Tibetan Buddhism. Nyingmapa combines all three forms of Buddhism: 'Hinayana to improve one's character, Mahayana to think about others, and Vajrayana to follow a short cut to spiritual liberation' (Zangbu and Klatzel, 2011). Guru Rinpoche was believed to have brought Buddhism to Khumbu and Tibet in the 8th century (Zangbu and Klatzel, 2011). In so doing he created numerous *beyuls* in the Himalayan region and hidden numerous treasures (*ters*) for Sherpas to discover in times of need (Klatzel, 2009). Sherpas strictly follow certain sets of rules in the *beyul*, such as not harming or killing any living things, respecting local spirits dwelling in the natural landscape, refraining from violence and negative thoughts, and keeping a peaceful and compassionate mind (Spoon, 2010). Therefore, Sherpas place Guru Rinpoche at a status near to the Buddha himself.

Like other Buddhists, Sherpa people believe in the reincarnation of various states of positive and negative being depending on the amount of sin or merit accumulated in one's lifetime

(Ortner, 1989). Accumulating sufficient merit and good deeds help one's chance of being reborn into a higher state of being and ultimately escaping the cycle of life, death, and rebirth. In addition to Buddhist concepts of merit, rebirth, and compassion, Nyingmapa is focused more exclusively on a fusion of Tantrism and Tibetan Bon practices which involves animistic and shamanistic rituals of protection, veneration, and exorcism (Ortner, 1989). Sherpas see the world as a place occupied by a variety of supernatural beings, including spirits (both good and bad) and deities who reside in trees, soil, water, mountains, and rocks. Offerings must be made to please the deities and spirits and to receive their protection. The spiritual values of the Sherpa are therefore closely associated with the natural landscape, and taboos related to the natural environment lead to sustainable practices, such as not cutting trees in certain areas, avoiding the contamination of soil and water sources, and not killing animals (Sherpa, 2008).

During our fieldwork, interviews conducted with local residents affirmed that Sherpas' understanding of the sacred landscape revolve around the *beyul*. Local residents were familiar with the story of the establishment and rediscovery of the *beyul*. Sherpas identify the entire Khumbu region, including the SNP and its peripheral areas in the lower land, as *beyul Khumbu*, one of the many sacred valleys across the Himalayan mountains. Other familiar *beyul* elsewhere in the Himalaya include Rolwaling, Langtang, Yolmo, and the restricted area of Khenpalung east of Khumbu (i.e. tourists are not permitted to enter). Some Sherpas were also familiar with *beyul* in Tibet, India, Bhutan, and China. According to one elder, a *beyul* can only be opened to 'the right people at the right time'. 'Right people' in this statement refers to those who achieve a certain level of spiritual development, while 'right time' refers to times of unrest when Tibetan religion is under threat from outside forces.

The concept of *beyul* goes beyond the sacred physical landscape and embodies abstract meanings too. One lama at the Pangboche Monastery explained that the *beyul* is both a physical landscape and a spiritual realm. What people see is the physical form of *beyul*, including mountains, forests and lakes, but the imaginary level of *beyul* can only be accessed through deep meditation and a higher level of spiritual attainment. Another lama, also from Pangboche Monastery, echoed a similar sentiment, stating that the *beyul* is both a physical domain and a metaphysical territory. He stated that the most powerful *beyul* exists only in highly enlightened minds, and in order to open up the spiritual *beyul*, people should 'always learn to keep a peaceful mind and practice good deeds for a purer heart'. In line with this thinking, the Khumbu residents believe that the *beyul* represents a utopian world where people's hearts and spirits are 'pure'. Avoidance of negative activities and thoughts, such as violence, killing, disturbing the land and wildlife, polluting the water and soil, and disrespecting human beings and spirits, is essential to maintain this 'purity'. The *beyul* was understood as 'a powerful sacred place',as it could 'purify people's soul[s]' and 'eliminate their sins'. Most Sherpa residents were well aware of the codes of conduct in *beyul* and considered it a precious space which needs to be carefully preserved. A *beyul* can be destroyed by 'negative forces', referring to conflicts among people and harmful thoughts in people's minds. Supernatural powers in the *beyul* would not protect people if they became 'selfish', 'uncompassionate', 'hateful', and 'did not cooperate with each other'. Many residents believed that when the guardian spirits in the land, water, and forests leave due to disapproved-of thoughts and behaviours, the *beyul* can be desecrated and lose its sacred power.

The Composition of Khumbu's Sacred Landscapes

Building on the concept of *beyul*, Sherpas told us in the interviews of three types of sacred landscape in the Khumbu region: natural, cultural and spiritual.

Natural features

Mountains were found to be the most recognized and significant physical feature among the sacred landscapes in Khumbu. Mt Khumbila and Mt Everest were the two most frequently mentioned sacred mountains (referred to 31 times and 27 times, respectively; see Table 5.1).

Table 5.1. Sacred mountains and their associated deities identified by Khumbu residents.

Sacred mountain identified by khumbu residents	Associated spiritual figure	Frequency of references
Khumbila	Khumbila	31
Everest	Jomo Miyo Lang Sangma	27
Taboche/Tawache	Tawuche	7
Amadablam	Shi Chu, Khumbila's gamekeeper	6
Thamserku	Tamosermu, Khumbila's wife	4
Kusum Khangaru	'Three snow-white gods'	3
Kangtega	No specific mention	3

Among all the sacred mountains in the region, Mt. Khumbila received a much higher level of respect and awe given its central geographical location and its residing deity Khumbu-Yul-Lha or Khumbila. However, our interviews revealed a lack of consistent knowledge among the residents as to why these mountains were sacred. A few residents could name the residing deities on these mountains, but many others were unsure. Seven residents stated that all mountains in Khumbu are sacred, or at least used to be. The loss of sacred status of some mountains was attributed to commercial mountaineering, which had caused the deities to leave the mountains. Geographic proximity appeared to influence residents' familiarity with the sacred values attached to these mountains. For example, Monjo and Phakding residents were more familiar with the deity residing on the adjacent mountain Kusum Khangkaru, but beyond these two villages no other Khumbu Sherpa we interviewed referred to this mountain as sacred.

Forests and water sources receive the highest level of protection in Khumbu, not only because they are indispensable resources for local livelihoods, but also because willful damage to these resources would offend the resident spirits. Monastic forests, in particular, were considered sacred territories. More than half of the residents we interviewed stated that forests near all monasteries (*gonpa*) were sacred (Table 5.2). Cutting trees in sacred forests was considered a serious offence to the spirits living in the woods by residents, as it would result in misfortunes and natural disasters such as landslides. Many residents also indicated that the sacred forests in Khumbu were shrinking in size, mostly due to the construction of new lodges and tourist shops. Permission from the local monastery for cutting trees is mandatory, and *pujas* must be held to venerate spirits for preventing negative impacts.

Groves of large trees close to springs and rivers were observed in several villages (e.g. Pangboche). These groves were sites of worship of the water spirit or deity *Lu* (*Lukhang*), believed to be bad tempered and easily disturbed. Local residents believed an unhappy *Lu* could abandon its home, causing the spring to dry up and bring suffering to people in the form of drought. *Lu* is also considered to be averse to contamination, demanding a high level of cleanliness. The majority of the residents stated that sources of water, especially those coming out of stone or emerging from the ground, signal a residing *Lu* and should be protected with great care. Worship of the *Lu* could be observed everywhere. For example, swing bridges were covered with layers of prayer flags, a gesture to seek blessings from the water spirit for peace and a good harvest. Similarly, the source of a spring was usually

Table 5.2. Frequency of references to sacred forests.

Name of sacred monastery forest	Frequency of references
All monastery forests	18
Tengbohe	7
Khumjung	5
Pangboche	4
Thame	2
Phortse	2
Phakding	1

surrounded by a low stone wall wrapped with white *khata* (a ceremonial scarf of goodwill and honour). Some private households designated a special stone shrine at the corner of their house for the deity *Lu*, as it could influence the family's health and luck.

Cultural features

Six types of sacred cultural feature were observed during our fieldwork and were also identified during interviews with residents: *gonpa* (monasteries), *stupa* or *chorten* (dome-shaped shrines), *kani* (entrance gates), manistones (stone plates inscribed with religious mantras), mani walls (linear alignment of manistones forming as wall), *chokhor* (prayer wheels), and prayer flags and poles.

A monastery, or *gonpa*, is an important sacred structure where veneration, religious education, and festivals take place. The location of each monastery is carefully selected – it is usually situated at higher ground, separated from residential and commercial areas, and marked by sacred features around the monastery building and its periphery. Among the monasteries in the Khumbu region, those in the villages of Tengboche, Khumjung, and Thame were identified as the three most important religious institutions and thus mentioned far more frequently by the residents. Many villagers narrated stories of the establishment of these monasteries as directly associated with Buddhist saints and legendary Buddhist leaders. The three monasteries mentioned above host two of the biggest religious festivals in Khumbu: *Mani Rimdu* and *Dumji*. *Mani Rimdu* is a sacred ceremony held over several days in autumn (usually in November) and consists of ritualistic masked dances, worship and prayers for empowerment, the welfare of humanity, and the triumph of Buddhism. *Dumji* is celebrated in July to honour the birth anniversary of *Guru Rinpoche*, or Sangwa Dorji, who the Sherpas believe was born on a lotus flower. In general, all residents we interviewed affirmed all of Khumbu's monasteries as being very sacred. However, the reverence of the monasteries was influenced largely due to residents' personal experience, engagement with, and attachment to, the monastic institutions.

We counted approximately 100 *stupa* or dome-shaped shrines along the main trekking route to Everest Base Camp (EBC). These stone memorials, also called *chorten*, varied in shape and size. Some were 'object stupa', which contain sacred scripts or relics of Buddhist saints. The majority of *stupas* in Khumbu, however, were 'memorial stupa', which are constructed to commemorate deceased lamas or to gain *sonam* (merit). Typically, a *stupa* is built with Buddha eyes on four sides to prevent the evil spirits from entering the village. Walking around a stupa in quiet contemplation is one of the daily religious practices Sherpas engage with as a way to venerate Buddhist deities and accumulate merit for personal spiritual development. *Stupas* also function as geographical landmarks in Khumbu. They are more concentrated near villages with a denser population such as Namche, Khumjung, and Pangboche. At least 23 memorial *stupas* were dedicated to foreign climbers and Sherpas who died in mountaineering expeditions. These memorial *stupas* were scattered along major trekking routes, but mostly concentrated near Lobuche, a village most trekkers must pass by on their way to EBC.

Another cultural feature that serves as a geographical landmark is the *kani*, a small gateway built at the entrance to a village marking its outer boundary. With its ceiling and walls delicately painted with religious figures, mantras, and mandalas, a *kani* serves as a protective feature to stop bad spirits that can follow a person when entering a village. Sherpas seek blessings by turning the prayer wheels inside the *kani* every time they pass through the gateway. We counted 12 *kani* gates in Khumbu, beyond Jorsalle where the SNP entrance is located. At the park entrance, a recently renovated *kani* stands along with huge mani walls (discussed below). On the bronze plate nearby, trekkers are warned that they are about to enter a sacred territory. The plate also provides information to tourists about codes of conduct, stipulating appropriate attitudes and behaviours.

Three other types of cultural features found in the sacred landscape of Khumbu reflect locals' veneration for supernatural power. First, a mani wall is an alignment of mani stones engraved with the sacred invocation 'Om Mani Padme Hum' which is believed to invoke spiritual power from Buddhist deities. They are usually found near major monasteries and big *stupa* or piled in front of households. Approximately 190 mani walls and big mani stone piles were recorded along the trekking trails. Carving (hand chiselled) mani stones or building mani walls are highly regarded and important contributions Sherpas make to their village, as it believed to help them accrue additional merit during their lifetime. Even foreigners occasionally make such gestures, as seen near Khumjung School where a mani wall was constructed with help from a South Korean mountaineering organization.

Second, prayer wheels can be built as an individual feature or aligned with many smaller wheels and mounted on the outer walls of a monastery. Inside the prayer wheels are rolls of religious scripts which contain sacred invocation of prayers. We counted about 65 big prayer wheels along the main trekking route to the EBC. These are mostly concentrated in and around local monasteries. In some villages, local residents had built a separate communal building to shelter these giant prayer wheels.

Third, the cultural objects most commonly seen in Khumbu are prayer flags in rectangular shapes hanging along strings and poles. Prayer flags are everywhere – along mountain ridges and on top of monasteries, households, *stupas*, trees and fields. Rows and rows of prayer flags can also be found hanging between rocks on steep slopes and along bridge railings. The prayer flags are offered as a way to appease resident spirits and seek safe passage. Sherpas believe that as the wind takes the powerful mantras written on these prayer flags up to the sky, goodwill can be spread and bring fortune to all beings. More than 660 sites with prayer flags and poles were counted during our fieldwork.

In addition to the above-mentioned religio-cultural features, which are mostly communal or public, visual manifestations of the Sherpa spiritual beliefs can be observed at a micro scale, namely inside and just outside a Sherpa home. The structure, layout, and orientation of Sherpa houses embodies religious symbolism and reflects local religious values. The family *sangkhang* or *lhasu* is a small fireplace decorated with small bundles of bamboo and prayer flags located outside the house for good fortune. Inside the house, most Sherpa families have a separate room for a *Lhang* (i.e. a private chapel). It is usually the most decorated room in the house and typically contains statues of religious figures, photographs of high lamas, sacred texts, and ritual objects. This is also where most daily religious routines are performed. Through chanting, filling seven water bowls, and lighting butter lamps on the altar ritually, Sherpas interact with the spiritual realms in their private space. In the Khumbu villages, Sherpas construct their homes facing away from the sacred mountains and springs so that they do not disturb the resident deities (Sherpa, 2008). This pattern is most evident in Khumjung, Phortse, and Thame, where the buildings are all oriented away from Mt Khumbila. Elder Sherpas indicate that the structure of a Sherpa house embodies religious meanings: short door frames and small windows are built to prevent bad spirits from entering the houses.

Spiritual figures

Our interviews with 33 resident Sherpas revealed five types of spiritual figure: Buddhist deities and saints, *yul-lha* (mountain deities), *lu* (water spirits), *sadak* (soil spirits), and reincarnated high lamas. While Buddhist deities, saints, and mountain deities were the most identified spiritual figures, some Sherpa elders revealed greater knowledge of local spirits and reincarnated high lamas. Older Sherpas, especially those over 50 and who had obtained a monastic education, demonstrated greater knowledge of Buddhist deities and saints, while younger Sherpas were less familiar with the legends and stories related to each deity.

Of all the deities, the Buddhist deity *Padmasambhava* (or Guru Rinpoche, meaning lotus-born) is the most important being, being at a status just below the Buddha himself. Sherpas honour him as the founder of Tibetan Buddhism and the creator of the sacred *beyul Khumbu*. Phakpa Cherenzig, the deity of compassion, was another frequently identified spiritual figure. Mani Rimdu, the biggest religious festival in Khumbu, is dedicated to the worship of Phakpa Cherenzig. At this event, monks perform sacred dances (*chham*) to show different manifestations of Cherenzig, and his blessings are put on *rilbu* (or *rimdu*, small red pills) and distributed to everyone in attendance; the red pill denotes long life, so the one receiving the blessed pill will live a long life. The knowledge of Buddhist deities and saints varied among different generations of Sherpa.

Yul-lha are believed to be associated with the indigenous Bon religion that was present before the arrival of Buddhism in the Himalayan region. The most frequent reference (31 times) was made to the mountain deity Khumbu-Yul-Lha, or Khumbila, the county deity of Khumbu who resides on the mountain named after him. Khumbila was believed to be a 'bad spirit' before Guru Rinpoche converted him into a good warrior and commended him to protect the people and land in Khumbu. Climbing Mt Khumbila is strictly prohibited, as this can enrage the deity and lead to serious consequences such as avalanches, earthquakes, droughts, and landslides. The villages of Khumjung and Khunde are located at the base of this mountain, and many Sherpa households in the villages conduct *puja* (ritual ceremonies) and give daily offerings to Khumbila in addition to invoking its spirit during various religious festivals. *Jomo Miyo Lang Sangma*, the female deity who resides on Mt Everest (or *Jomolungma* in Tibetan) was mentioned by 27 residents. She is one of the five sisters of long life in traditional Tibetan mythology who gives food and wealth. Although the deity is widely recognized as an important spiritual figure in Khumbu, Mt Everest itself has less religious significance for Sherpas. Ritual ceremonies usually take place before a climbing expedition but not on a daily, regular basis. Several residents indicated that there are *yul-lhas* on every mountain peak in Khumbu but that their individual names are not specific (they are known by multiple names). Sherpas usually mentioned the sacred mountains near their home villages first because they were more familiar with the associated deities and the names of the mountains. However, Khumbila and Jomolungma were two exceptions, being referred to by almost all the residents we interviewed.

In addition to the mountain gods, local Sherpas also acknowledged the presence of water spirits (*lu*) and soil spirits (*sadak*). These spirits are believed to live in the water, trees, rocks, and earth. Their habitats are sensitive to contamination and defilement, and therefore keeping these areas clean was deemed crucial. Veneration of these minor spirits usually occurs at the origin of water sources or inside and just outside individual households. Although most residents were aware of these spirits, the exact dwelling locations of these spirits were unknown to many with the exception of some elderly Sherpas. Some residents were able to identify several *lu* spots in their villages during interviews but were unfamiliar with *lu* in other villages.

Reincarnated high lamas were the least identified spiritual figures by local Sherpas, probably because they are not supernatural *per se*. In the Khumbu region, some high lamas and respected Buddhist teachers who contributed significantly to the establishment and spread of Tibetan Buddhism are believed to have reincarnated or achieved enlightenment after their death, such as Lama Sangwa Dorje, who promoted Tibetan Buddhism and built monasteries and communities throughout Khumbu more than 350 years ago. It is a common practice that paintings and photos of reincarnated high lamas are displayed in Sherpa homes.

Sacred Landscapes and Tourism: Harmony or Discord?

When local residents were asked if tourism development had somehow influenced their beliefs associated with Khumbu's sacred landscape, most noted three specific influences: 1) altered conception of the *beyul*; 2) shifting spiritual values; and 3) associated changes in religious practices (see Mu *et al.*, 2019).

Interviews with local Sherpa indicated that although tourism changed their religious practices and spiritual values (see below), most of them believed that the sacred *beyul* still held a very special status. The majority of those interviewed believed that Khumbu continues to remain a sacred place in spite of increasing commercialization and westernization. According to their belief, a sacred *beyul* can exist in modern society as long as Sherpas make effective efforts to preserve their culture and religion. Some viewed the transformation of Sherpa culture as inevitable and uncontrollable, and attributed it to Buddhism's principle of 'impermanence' – the idea that nothing remains the same forever. However, local residents also emphasized the importance of local trekking guides informing and educating tourists about Sherpa culture and religious values so they would behave in a culturally acceptable manner in the sacred *beyul*. For Sherpas who worked in dangerous and physically demanding occupations such as high mountain guides and porters, their interactions with the *beyul* were somewhat different. Given the travel demands of their career, mountain guides and porters are not able to participate in some daily ritual routines practised by many Sherpas. For example, some Sherpas have to travel around Nepal working for foreign expedition teams in high mountains throughout the year. However, interviews with some mountaineering guides indicated that they have a much deeper understanding of the *beyul* than other local residents.

Sherpa perceptions of the spiritual aspects of the physical landscape came directly from intensive interactions with nature itself. As one of the guides stated:

> [Before an expedition], I always encourage my clients to go visit a local gonpa with me and receive blessings from lamas. We also perform a simple puja and make offerings to the mountain gods before every important ascent. We ask their permission and protection by saying things like 'please allow us to approach you, we are all good and humble people ...' If something is wrong [during the climb], like ... a sudden weather change, or an unusual obstacle appears, we give up and go back to camp. That means the gods don't want you to climb today! You will get to the top at a right time, but not today.

Similarly, tourism's influence on Sherpa spiritual values have occurred in three ways (Mu *et al.*, 2019). First, tourism has changed the ways in which mountain deities are worshipped. In the past, Sherpas used to pray to mountain deities asking for a good harvest and good weather. Today, their prayers include asking for more businesses, for the safety of family members in mountaineering expeditions, and the well-being of their children living outside Khumbu (mostly Kathmandu). One resident stated that with jobs in the trekking and mountaineering tourism industry being very risky, the importance of prayers has been magnified. As she stated:

> Every morning, I burn incense and butter at the *lhasu* (private shrine) in our backyard; it's a respect for the land deities. Then I go to the *lhapso* [village shrine] to burn some juniper as offering to the gods in mountains. At night I pray in [the] *lhang* [prayer space at home] and make offerings to Buddha ... [We] make offerings to mountain gods because they easily get angry and cause avalanches. My husband works as a porter in April and May. During that time, I pray a lot for his safety in the mountains. Every time he returns safe, we go to *gompa* together and thank all the deities for protecting him.

Sherpa perceptions about the sacredness of specific mountains is changing too. Mt Everest used to be perceived as the dwelling of the goddess *Jomo Miyo Lang Sangma*, but it was not held in high in esteem as Mt Khumbila, the dwelling place of Khumbila. As such, climbing a sacred mountain like Everest and other expedition peaks in Khumbu was tolerated and not seen as denigrating the deities that reside in them. While the goddess *Jomo Miyo Lang Sangma*, as noted earlier, was traditionally seen as being responsible for food and wealth, today she is linked to prosperity in the tourism industry. Many Sherpa families, especially those who are involved in tourism, now pray to her for more tourists. Mountaineering guides pray for their safety on Mt Everest by conducting *pujas* at the base camp before climbing higher up. As mountaineering tourism expanded and flourished in Khumbu, climbing sacred mountains like Everest was negotiated as morally and spiritually acceptable. According to the interviewees, contemporary materialistic values did not clash with Sherpa spiritual values; tourism provided new perspectives and meanings to existing cultural traditions and social norms. For example, a large tourist presence at festivals such as Mani Rimdu was seen as contributing to enhanced self-awareness among the Sherpas of their culture and values. Even stories narrated during the festivals incorporate comical (sometimes educational) encounters between the Sherpa and the tourist, suggesting that hospitality and tourism are now integral to Sherpa culture.

Second, traditionally, the relationship between the Sherpa and the land was a symbiotic one. Strong religious beliefs and place-based spiritual values constrained and regulated residents' use of the land and its natural resources. In a sacred *beyul*, humans, landscapes, and supernatural entities are interconnected with the *beyul*, symbolizing the 'umbrella that links humans to each other and the land' (Spoon, 2010,

p. 95). Today, the sacred landscapes of Khumbu have been symbolized in a somewhat different way, as monetized commodities (Skog, 2017). The natural and cultural landscapes in Khumbu are now seen as products, sold to foreign tourists, as the Pangbochelama stated: '... our religious festivals, traditional clothes and arts, everything can make money now!'

Third, as revenue generated from tourism has become a major source of income, Sherpa participation in religious activities has continued to decline. Time and resources are now mostly dedicated to the upkeep of tourism businesses rather than religious activities. Western (English) education is now preferred by most Sherpas, particularly by those who plan to get involved in tourism. The number of local Sherpas receiving a monastic education has dropped drastically. The inheritance of traditional language and religious knowledge is becoming more challenging as fewer Sherpas dedicate themselves in pursuing spiritual development. However, during the monsoon season when there are not many tourists in Khumbu villages, Sherpa engagement in religious activities is more frequent. For example, the *Dumchi* festival during the monsoon season provides an opportunity for Sherpa people outside of Khumbu to reunite with their extended families in the region. Although some of the traditional rituals and practices have been simplified and less formalized, Sherpa passion and devotion toward their religious and cultural heritage has remained strong.

Most residents considered tourism-introduced changes in Khumbu as more beneficial than harmful. Four positive outcomes of tourism were frequently referred to: greater accessibility to the world, improved infrastructure, rising economic prosperity, and educational improvements, such as better equipped schools and good quality teachers. However, many stated that the Sherpa people have become less religious owing to increased commercialization and westernization. For instance, prior to tourism, Namche and Pangboche used to be small settlements. Today, these villages are major tourist hubs, and the presence of Irish pubs, internet cafes, coffee shops, and western restaurants has introduced a completely different lifestyle to local Sherpa people and has reshaped their views of the world around them. Modern influences such as those observed in major tourist settlements like Namche will arguably erode Sherpa cultural and religious traditions, but research has also indicated that the Sherpa people have learned the art of careful negotiations between tradition and modernity (Nepal, 2015). Likewise, Sherpas' interest in continuing or maintaining cultural traditions is also extended to environmental protection efforts, for example, through collective actions geared toward waste management, environmental clean-up, trail maintenance, and forest management.

Conclusion

This chapter has looked at Sherpa perspectives of sacred landscapes in Khumbu, a major trekking and mountaineering destination. It examined the composition of sacred landscapes as manifested in Sherpa conceptions of the interconnections between the natural, the cultural, and the spiritual domains of the Khumbu *beyul*. The Sherpas are devout Buddhists who believe in the natural and supernatural worlds as defining their daily religious practices and rituals, and their cultural traditions and values. Somewhat alluded to, but not explicitly discussed in this chapter, is the idea that Khumbu's sacred landscape has a hierarchical structure with multiple elements present within it. These elements can range from the abstract and intangible in the supernatural realm to concrete features in the physical and cultural landscape as manifested in specific rituals and objects which serve to reflect and reinforce spiritual faith.

Khumbu has evolved from a remote region, once dependent on a subsistence economy, to one that is cosmopolitan and oriented towards international tourism. Arguably, Khumbu's adventure tourism hinges on Sherpas' deeply held spiritual beliefs and cultural practices that guide their interactions with tourists, as the Sherpas lead tourists through dangerous topographies and save tourist lives during disasters, often at the cost of their own lives. Many Sherpas involved in Khumbu's tourism industry make genuine attempts to impart these cultural practices to the tourists they encounter. However, the transactional nature of exchanges between the Sherpas (as hosts) and the foreigners (as guests) has also meant that tourism has influenced how the Sherpas view their own place in the physical and metaphysical world. It is safe to conclude that tourism has not compromised Sherpa spiritual beliefs and cultural practices, but has rather reordered, reproduced, and realigned these beliefs and practices to adapt to a changing world.

References

Adams, V. (1992) Tourism and Sherpas, Nepal: reconstruction of reciprocity. *Annals of Tourism Research* 19(3), 534–554.

Bremer, T.S. (2006) Sacred spaces and tourist places. In: Timothy, D.J. and Olsen, D.H. (eds) *Tourism Religion and Spiritual Journeys*. Routledge, Abingdon, UK, pp. 25–35.

Byers, A. (2005) Contemporary human impacts on alpine ecosystems in the Sagarmatha (Mt. Everest) National Park, Khumbu, Nepal. *Annals of the Association of American Geographers* 95(1), 112–140. DOI: 10.1111/j.1467-8306.2005.00452.x.

Cosgrove, D.E. (1998) *Social formation and symbolic landscape*. University of Wisconsin Press, Madison, Wisconsin.

Department of National Parks and Wildlife Conservation (2017) *State of Conservation Report: Sagarmatha National Park (Nepal) (N 120))*. Ministry of Forests and Soil Conservation, Kathmandu, Nepal.

Dudley, N., Higgins-Zogib, L. and Mansourian, S. (2009) The links between protected areas, faiths, and sacred natural sites. *Conservation Biology* 23(3), 568–577. DOI: 10.1111/j.1523-1739.2009.01201.x.

Horster, M. (2010) Religious landscape and sacred ground: relationships between space and cult in the Greek world. *Revue de l'histoire des religions* 4(227), 435–458. DOI: 10.4000/rhr.7661.

Hubert, J. (1997) Sacred beliefs and beliefs of sacredness. In: Carmichael, D.L., Hubert, J., Reeves, B. and Schanche, A. (eds) *Sacred Sites, Sacred Places*. Routledge, Abingdon, UK, pp. 9–19.

Jazeel, T. (2018) Building sacred modernity: Buddhism, secularism and a geography of 'religion'in southern Sri Lanka. In: Bartoilini, N., MacKian, S. and Pile, S. (eds) *Spaces of Spirituality*. Routledge, London, pp. 68–80.

Klatzel, F. (2009) *Gaiety of Spirit: The Sherpas of Everest*. Mera Publications, Kathmandu.

Knudsen, D.C., Soper, A.K. and Metro-Roland, M.M. (2008) Landscape, tourism, and meaning: an introduction. In: Knudsen, D.C., Soper, A.K., Metro-Roland, M.M. and Greer, C.E. (eds) *Landscape, Tourism, and Meaning*. Ashgate Publishing, Abingdon, UK, pp. 1–8.

Knudsen, D.C. and Greer, C.E. (2008) Heritage tourism, heritage landscapes and wilderness preservation: the case of National Park Thy. *Journal of Heritage Tourism* 3(1), 18–35. DOI: 10.1080/1743873X.2008.9701248.

Lane, B.C. (2002) *Landscapes of the Sacred: Geography and Narrative in American Spirituality*. The John Hopkins University Press, Baltimore and London.

Levi, D. and Kocher, S. (2013) Perception of sacredness at heritage religious sites. *Environment and Behavior* 45(7), 912–930. DOI: 10.1177/0013916512445803.

Luger, K. (2000) *Kids of Khumbu, Sherpa Youth on the Modernity Trail*. Mandala Book Point, Kathmandu, Nepal.

Ministry of Culture, Tourism, and Civil Aviation (MOTCA) (2017) *Nepal Tourism Statistics 2016*. MOTCA, Kathmandu, Nepal.

Mu, Y., Nepal, S.K. and Lai, P.-H. (2019) Tourism and sacred landscape in Sagarmatha (Mt. Everest) National Park, Nepal. *Tourism Geographies* 21(3), 442–459. DOI: 10.1080/14616688.2018.1558454.

Mücher, C.A., Klijn, J.A., Wascher, D.M. and Schaminée, J.H.J. (2010) A new European landscape classification (LANMAP): a transparent, flexible and user-oriented methodology to distinguish landscapes. *Ecological Indicators* 10(1), 87–103. DOI: 10.1016/j.ecolind.2009.03.018.

Nepal, S.K. (2003a) Trail impacts in Sagarmatha (MT. Everest) National Park, Nepal: a logistic regression analysis. *Environmental management* 32(3), 312–321. DOI: 10.1007/s00267-003-0049-7.

Nepal, S.K. (2003b) *Tourism and the Environment: Perspectives from the Nepal Himalaya*. Studien Verlag, Innsbruck & Himal Books, Lazimpat, Kathmandu.

Nepal, S.K. (2005) Tourism and remote mountain settlements: spatial and temporal development of tourist infrastructure in the MT Everest region, Nepal. *Tourism Geographies* 7(2), 205–227. DOI: 10.1080/14616680500072471.

Nepal, S.K. (2015) Irish pubs and DREAM cafes: tourism, tradition and modernity in Nepal's Khumbu (Everest) region. *Tourism Recreation Research* 40(2), 248–261. DOI: 10.1080/02508281.2015.1047625.

Nepal, S.K. (2016) Tourism and change in Nepal's Mt. Everest Region. In: Richins, H. and Hull, J.S. (eds) *Mountain Tourism – Experiences, Communities, Environments and Sustainable Futures*. CAB International, Wallingford, UK, pp. 285–294.

Nepal, S.K. and Nepal, S.A. (2004) Visitor impacts on trails in the Sagarmatha (Mt. Everest) National Park, Nepal. *AMBIO: A Journal of the Human Environment* 33(6), 334–340.

Nepal, S.K., Kohler, T. and Banzhaf, B.R. (2002) *Great Himalaya: Tourism and the Dynamics of Change in Nepal*. Swiss Foundation for Alpine Research, Zurichin collaboration with the Centre for Development and Environment, University of Berne, Berne, Switzerland.

O'Hare, D. (1997) Interpreting the cultural landscape for tourism development. *Urban Design International* 2(1), 33–54. DOI: 10.1057/udi.1997.5.

Ormsby, A. (2013) Analysis of local attitudes toward the sacred groves of Meghalaya and Karnataka, India. *Conservation and Society* 11(2), 187–197. DOI: 10.4103/0972-4923.115722.

Ortner, S.B. (1989) *High Religion: A Cultural and Political History of Sherpa Buddhism*. Princeton University Press, Princeton, USA.

Perriam, G. (2015) Sacred spaces, healing places: therapeutic landscapes of spiritual significance. *Journal of Medical Humanities* 36(1), 19–33. DOI: 10.1007/s10912-014-9318-0.

Setten, G. (2006) Fusion or exclusion? Reflections on conceptual practices of landscape and place in human geography. *Norwegian Journal of Geography* 60(1), 32–45.

Sherpa, L.N. (2008) *Through a Sherpa Window: Illustrated Guide to Traditional Sherpa Culture*. Vajra Publications, Kathmandu.

Spoon, J. (2010) Tourism meets the sacred: Khumbu Sherpa place-based spiritual values in Sagarmatha (Mount Everest) National Park and buffer zone, Nepal. In: Verschuuren, B., Wild, R., McNeely, J.A. and Oviedo, G. (eds) *Sacred Natural Sites: Preserving Nature and Culture*. Earthscan, London, pp. 87–97.

Spoon, J. and Sherpa, L.N. (2008) Beyul Khumbu: the Sherpa and Sagarmatha (Mount Everest) National Park and buffer zone, Nepal. In: Mallarach, J.M. (ed.) *Protected Landscapes and Cultural and Spiritual Values*. Kasparek Verlag, Heidelberg, pp. 68–79.

Skog, L.A. (2017) *Khumbi yullha* and the *Beyul* : sacred space and the cultural politics of religion in Khumbu, Nepal. *Annals of the American Association of Geographers* 107(2), 546–554. DOI: 10.1080/24694452.2016.1210498.

Stevens, S.F. (1993) Tourism, change, and continuity in the Mount Everest region, Nepal. *Geographical Review* 83(4), 410–427. DOI: 10.2307/215823.

Stokols, D. (1990) Instrumental and spiritual views of people-environment relations. *American Psychologist* 45(5), 641–646. DOI: 10.1037/0003-066X.45.5.641.

Stump, R.W. (2008) *The Geography of Religion: Faith, Place, and Space*. Rowman & Littlefield, Lanham, Maryland.

Terkenli, T.S. (2004) Tourism and landscape. In: Lew, A.A., Hall, C.M. and Williams, A.M. (eds) *A Companion to Tourism*. Blackwell Publishing, Oxford, UK, pp. 337–348.

Tilley, C. and Cameron-Daum, K. (2017) *An Anthropology of Landscape*. UCL Press, London, UK.

Zangbu, N.T. and Klatzel, F. (2011) *Stories and Customs of the Sherpas*. Mera Publications, Kathmandu.

6 Religious Tourism and Environmental Conservation in Lumbini, the Birthplace of Lord Buddha, World Heritage Site, Nepal

Anouk Lafortune-Bernard[1]*, Rajendra N. Suwal[2], Kai Weise[3] and Robin A.E. Coningham[4]

[1]Department of Archaeology, Durham University, United Kingdom; [2]WWF Nepal; [3]ICOMOS Nepal; [4]UNESCO Chair, Department of Archaeology, Durham University, United Kingdom

Introduction

Buddhist texts and iconography demonstrate a pervasive tradition of the representation of natural environments, notably in relation to the life of the historical Buddha, Siddhartha. Not only did his past lives include numerous animal forms, but his own stories and parables often included animal protagonists. Trees were particularly important in these stories and were associated with all four major events in his life: his birth in the Garden of Lumbini; his enlightenment under the Bodhi Tree in Bodh Gaya; his first teaching in the deer park of Sarnath; and his passing away in Kushinagara. There has also been a long tradition of the use of tree shrines in Buddhist ritual practices, being present in both iconographic representations and historical texts, and continuing to be venerated to the present day in Bodh Gaya in India and Anuradhapura in Sri Lanka.

In the context of increasing environmental awareness, a rich literature has developed to focus on the links between Buddhism and ecology using historical, textual, and archaeological evidence alongside examples from contemporary Buddhist practices (e.g. Badiner, 1990; Shaw, 2016; Chapple, 2017). In relation to the development of environmental practices, the protection of water sources, forests, and animals have been of particular concern. Shaw (2016, pp. 529–531) has notably discussed the archaeological evidence for early forms of environmental control related to the regulated use and management of these natural resources developed by Buddhist monasteries from at least the 3rd century BCE.

According to textual traditions, the historical Buddha was born in the Garden of Lumbini as his mother, Queen Mayadevi, rested during her journey from her husband's city, Kapilavastu, to her father's home in Devadaha (IUCN, UNESCO Kathmandu and OCHSPA, 2013, p. 33). While the description of the garden varies extensively in different sources, the main components of the birth story are consistent: Mayadevi bathed in the Puskarni Pond and gave birth to her son, holding on to the branches of a Sal tree. The location of Lumbini in the modern Nepal's Western Terai was re-discovered in 1896 following the

*Corresponding author: a.lafortune91@gmail.com

discovery of an inscription on a pillar recording the pilgrimage of the Emperor Asoka in the 3rd century BCE (IUCN, UNESCO Kathmandu and OCHSPA, 2013, p. 33).

Today, Lumbini is an important pilgrimage site and a tourist attraction in Nepal. Over the last decade, visitor numbers have increased rapidly from 588,000 in 2011 to 1.5 million annual visitors in 2017. The site particularly attracts a domestic market and visitors from other South, South-east Asian, and East Asian countries. Pilgrims and tourists visit several key features of both sacred and archaeological value within the Sacred Garden, including the Maya Devi Temple, Puskarni Pond, Asoka's inscription on his Pillar, *stupas*, and monastic areas – all of which are closely associated with the development of the site as a place of worship and Buddhist pilgrimage from the 6th century BCE onwards (Coningham *et al.*, 2013).

Since 1978, the development of the site and its immediate surroundings has been guided by the Lumbini Master Plan (LMP) covering a five-by five-mile area that is divided into a central one by a three-mile Project Area and surrounding buffer zones. The implementation of the LMP has completely transformed the site's immediate surroundings, characterized previously by expanses of paddy fields with clusters of small agricultural villages that had 'changed little in the last 2500 years' (Allchin and Matsushita, 1969, p. 10). While the LMP included major infrastructure for national and international pilgrimage and tourism activities, it also mirrored the importance of the natural environment in the life of the Buddha. As a result, gardens, forested areas, and water bodies formed central elements in the design. The LMP also recommended the strict control of activities within the five by five-mile area around the Sacred Garden in order to preserve the sanctity, quietness, and beauty of the area of Lumbini and create a protected sacred landscape.

This chapter discusses how the LMP integrated the surrounding environment in the site development and the challenges that site managers have faced in implementing this vision and balancing its social, economic, and environmental impacts. With the LMP implementation in its final stages in 2019, the chapter begins by reflecting on the design and its use of the natural environment and landscaping components in its conception of the pilgrim experience. The second section critically reviews the implementation of the plan in comparison with this original design and considers its impact on local dynamics, particularly that of the Project Area and its surrounding communities. In the rapidly changing context of increasing visitor numbers and accelerated development within the Greater Lumbini Area, driven notably by international and national investments in regional infrastructure, this chapter concludes by discussing the current and future challenges linked to the management of Lumbini beyond the completion of the LMP and the sustainable future of its wider region.

Conception of the Lumbini's Master Plan

The visit of United Nations Secretary-General U Thant to Lumbini in 1967 is widely acknowledged as the turning point for the modern development of the site. At the time of U Thant's visit, Lumbini's surroundings remained primarily agricultural with small clusters of villages and a population of 1050 living within what is now the Project Area (KTU, 1978). There was limited infrastructure and facilities for visitors, and access to the site was challenging, with mud tracks interspersed with ferries (Dharmodaya Sabha, 1956; UNESCO, 2013). U Thant initiated a discussion with the Government of Nepal 'to develop Lumbini into an international pilgrimage and tourism centre' (UNESCO, 2013, p. 57) with support from the international community. Beyond the religious and spiritual motives, the socioeconomic development of the wider region was at the core of the project from its inception (UNESCO, 2013, p. 57).

After a decade of conceptualization and preparation, the LMP final design was approved by the Government of Nepal in 1978 (Fig. 6.1). The LMP conceptualized a pilgrim experience with visitors walking from the secular Cultural Zone, through the spiritual Monastic Zone, and finally entering the Sacred Zone, at the centre of which is the Garden of Lumbini with its archaeological remains (KTU, 1978). Within the first two zones the plan included practical infrastructure for pilgrims and tourists,

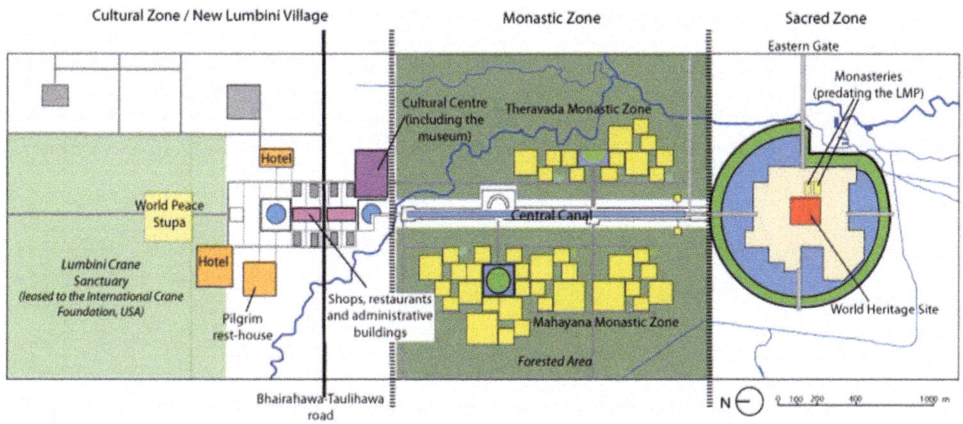

Fig. 6.1. A Sketch plan of the Lumbini Master Plan with three zones that visitors walk through from the Cultural and the Lumbini Village to the forested Monastic Zone and finally the Sacred Garden. The Lumbini Crane Sanctuary is a later addition to the original design with a lease for 100 hectares signed in 1994 between the Lumbini Development Trust (LDT) and the International Crane Foundation, USA. The sanctuary is currently managed by the Lumbini Crane Conservation Centre. (Anouk Lafortune-Bernard, adapted from KTU, 1978 and UNDP/UNESCO, 2013, p. 33).

notably accommodation, areas for restaurants and shops, a museum, and meditation centres and monasteries from different Buddhist traditions where pilgrims could interact with an international monastic community. Within the Sacred Zone, by contrast, the LMP prohibited 'any kind of constructions to preserve its archaeological value' (KTU, 1978, p. 20). As tree roots can cause significant damage to archaeological remains, the design for the Sacred Garden was primarily based on simple, non-intrusive, and reversible landscaping and walkways (KTU, 1978, pp. 19–20; Coningham and Acharya, 2013, p. 6). However, the LMP still placed significance on the creation of a naturalized environment around the Sacred Garden.

The main components of this naturalized environment outlined in the LMP included the reintroduction of forested areas, flower gardens, and the creation of new water bodies, the latter being designed to control annual flooding during the monsoon (KTU, 1978, pp. 19–20). These natural additions were to act as key components in the division of space connecting or separating key features and areas within the LMP. While the Sacred Garden, isolated by a circular levee, was to remain empty of any prominent features other than archaeological remains, the monastic zones on both sides of the central canal were to be reforested. These forested areas would reinforce the visual link formed by the central canal and walkways between the Cultural and the Sacred Zone, preparing 'the pilgrim for the culmination of experience' as they approach the sacred site (KTU, 1978, p. 61). The flower gardens, water bodies, and forested areas were therefore all designed to provide a natural setting around the Garden of Lumbini which closely reflected the various natural features described in Buddhist textual traditions.

In the LMP design, the visitor experience in Lumbini and the required facilities and services were all contained within the Project Area. To facilitate access to the site, the design included the development of transportation infrastructures in the wider region, notably the upgrade of the local airport at Bhairahawa to accommodate commercial domestic flights and the construction of a tarmac road linking Bhairahawa to Lumbini. The latter road then continued across to the Kapilbastu District where other major archaeological sites had been identified (UN, 1970). In the immediate area surrounding the Project Area, the LMP defined buffer zones to function as a transitional zone between the Project Area and activities in the wider region. Within these buffer zones, the plan outlined regulations to protect the site's setting. Production activities within the buffer zone of three by three-miles were to be limited to agriculture. Beyond this boundary, industrial production was to be prohibited within the entire five by five-mile restricted zone to preserve the climate, hydrology, and fauna and flora of the region. These planning regulations and the strict control of activities within the LMP area supported the successful inscription of the site on UNESCO's World Heritage list in 1997 based on Lumbini's archaeological, historical, and religious significance (ICOMOS, 1996).

Implementation of the Lumbini Master Plan

While the regional components (i.e. the Bhairahawa Airport and the Bhairahawa-Lumbini-Taulihawa road) have been completed and are now being upgraded with funding from the Asian Development Bank (UNDP/UNESCO, 2013, p. 27), the implementation of the other components is still ongoing after 40 years. The process has been affected by long delays, exponential increases in project costs, and other managerial issues, including limited partnerships and coordination between key stakeholders (UNESCO, 2013; UNDP/UNESCO, 2013). The responsibility for the completion of the Project Area lies with the site manager, the Lumbini Development Trust (LDT), under the 1985 Lumbini Development Act. The Department of Archaeology (Government of Nepal) shares jurisdiction of the Sacred Garden World Heritage Site as the national authority responsible for the protection of heritage. Ultimately, the buffer and restricted zones are under the responsibility of the local authority, formerly Village Development Committees that are now part of the Lumbini Cultural Municipality. The regulations outlined in the LMP to create a protective buffer around the Project Area have never been implemented in the former Village Development Committees nor in the Municipality and new plans have been proposed for their development. During the implementation phase, numerous changes have been made to the Project Area's original design, most notably additional constructions within or near the Sacred Zone, which constructions have been criticized for not abiding by the LMP planning principles and its strict building restrictions within the proximity of the Sacred Garden (WHC, 2004, p. 114).

Most of the key components in the central zone, including the monasteries and the central canal and walkways, were completed during the 1990s and early 2010s. The circular levee and central canal, for instance, were only completed in 2011. Over the same period, hotels and guesthouses opened just outside the eastern gate of the Sacred Garden within the agricultural buffer area rather than within the Cultural Zone. All these factors have affected visitors' movement and experience in Lumbini. Rather than following the original design with visitors progressing from the cultural to the spiritual to the sacred area, many visitors and pilgrims enter through this eastern gate (Fig. 6.1) and, until recently, only visited the Sacred Garden. A visitor survey conducted in 2013 indicated that most visitors to Lumbini spend very little time at the site, spending on average less than 30 minutes within the Sacred Garden, with 66% of visitors not staying overnight (Coningham and Acharya, 2013, pp. 50–60). Beyond the Sacred Garden, the Monastic Zone attracted a little over half of visitors coming to Lumbini (Coningham and Acharya, 2013, pp. 50–60). The visitors' routes have therefore tended to overlook the Cultural Zone, where the visitors' facilities were planned, and the forested areas within the Monastic Zone. Not only do these omissions minimize the 'culmination of experience' at the heart of the LMP design, they have also affected visitors' satisfaction, as they do not find within the Sacred Garden all the facilities, information, and 'natural setting with gardens and large trees' that they expect (UNESCO, 2013, p. 168). Ultimately, the limited time spent by visitors has also affected economic benefits for local communities, notably for the local guesthouses, restaurants, and shops located in the overlooked Cultural Zone and Municipality (Molesworth and Müller-Bölke, 2005; Pandey, 2007).

A UNESCO/Japanese-Funds-in-Trust project reviewed the LMP design and its implementation in 2013 to address these issues and integrate changes that have occurred on the site since 1970s, including the World Heritage listing of the Sacred Garden area and the growth in visitor numbers. The review provided suggestions to reintroduce the original vision concerning visitor experience and movement (Nishimura, 2013). With the completion of key components in the Central Zone and the implementation of suggested development in the Cultural Zone, including a newly built visitor centre, bus park, and other transportation services, an increasing number of visitors are now following the preferred LMP route, although their length of stay remains limited.

Balancing Environmental Conservation and Local Communities

Today, the naturalized environment of the Project Area is a combination of grassland (over 58%), forest plantations (40%), artificial water bodies, natural rivers, and wetlands (over 1%) (UNESCO, 2013, p. 147). These non-agricultural areas and new natural habitats introduced by the plan have attracted birds, reptiles, and other animals, including vulnerable and endangered species like the Nilgai 'blue bull', the largest antelope in South Asia, and the sarus crane, the tallest flying bird, closely associated with the Buddha's childhood (Fig. 6.2).

Many organizations have been involved in tree plantation programmes and environmental conservation projects in the Project Area, including the International Union for Conservation of Nature (IUCN) Nepal, the United Nations Development Programme (UNDP), and the World Wildlife Fund (WWF) Nepal (UNESCO, 2013). The International Crane Foundation has also a 50-year lease to manage and conserve an area of 100 hectares in the Cultural Zone as the Sarus Crane Sanctuary (Suwal and Bhuju, 2006). As the sarus crane has increasingly become 'a symbol of Lumbini and a flagship species for conservation in the region' (Baral, 2019, p. 176), the Crane Sanctuary has attracted growing interest in recent years from various local, national, and international organizations. Its ambition has been 'to make the Sanctuary a sustainable destination for Buddhist pilgrims to learn about the wildlife's central position in the historical ecosystem' (Ball, 2019, p. 197).

Fig. 6.2. Nesting sarus cranes in Lumbini (courtesy of R.N. Suwal, used with permission).

However, the development of Lumbini's naturalized environment and associated conservation efforts continue to face several challenges. While large-scale tree plantation programmes have been undertaken, with over 370,000 saplings planted in the early implementation phases, the forested areas envisioned in the LMP design have struggled to materialize on both sides of the central canal as a high proportion of saplings have not survived. The mixed results have been partially explained by the choices of species, including the introduction of non-native species and the development of monoculture blocks, due to the predominance of sisau trees in the plantation programmes, within which fungal diseases have more easily spread (Suwal and Bhuju, 2006, p. 94; Watson and Rai, 2019, pp. 166–167). In response to these issues, between 2012 and 2016 WWF Nepal, in partnership with LDT, planted over 600,000 saplings of different native species related to the Buddha's life. The re-creation of habitats and forests which reflect, protect and promote the diversity of the local ecosystems has therefore become a more central consideration in planting strategies and environmental conservation (Watson and Rai, 2019, p. 168).

Beyond the choices of species and plantation strategies, the competing uses of natural resources within the LMP between local residents, site managers, and other stakeholders has also affected the success of the tree plantation programmes (Molesworth and Müller-Bölke, 2005, pp. 204–205). The cutting of trees for firewood, logging, cattle grazing, waste dumping, and soil excavations inside the LMP have all been activities that have affected vegetation growth, the survival of saplings, and the overall forested setting within the LMP (UNESCO, 2013, p. 162). Moreover, harmful activities have also targeted local fauna, including fishing and poaching endangered animals – notably blue bulls and pythons – and the illegal trading of endangered bird species (UNESCO, 2013).

These challenges are not unique to Lumbini, and previous experiences and studies in the Western Terai and across the border in Uttar Pradesh have highlighted the determinant role of local communities – notably farmers' 'tolerant attitude', local traditions, and religious beliefs – for the protection of endangered species (Gopi Sundar, 2010, pp. 35–36; see also Shrestha, 1995). Community engagement, in the form of awareness-raising programmes but also in promoting 'the appropriate use of Lumbini, and shar[ing] the benefits from the conservation of Lumbini and the surrounding area', is therefore regarded as essential to deliver the naturalized environment envisioned in the LMP (Watson and Rai, 2019, p. 169).

At present, however, many among the local communities have perceived the LMP implementation activities as excluding them 'from the social development process' (Pandey, 2007, p. 15). Indeed, the LMP planning team had already recognized in 1978 that the LMP could have negative economic and social impacts on communities living within or just outside the Project Area. They specifically identified village resettlements, the loss of agricultural lands, and changes in traditional irrigation as possible negative effects on local communities (KTU, 1978). Land acquisition was also identified as a significant source of conflict between implementing agencies and local communities, especially in the cases of forced relocation, particularly due to what was perceived to be insufficient levels of compensation for land and houses (Molesworth and Müller-Bölke, 2005, pp. 194–195). Interviews conducted at later stages of the implementation of the LMP in the mid-1990s and early 2000s indicated continued resentment regarding these issues, particularly among more marginalized households and population groups living in the surrounding villages (Molesworth and Müller-Bölke, 2005; Pandey, 2007). Communities also reported that they lost access to natural resources, such as firewood and clay for houses, which they had collected freely prior to the LMP. Moreover, the growth of populations of wild boar and blue bull has affected agricultural activities outside the boundaries, as animals have strayed beyond the Project Area into the surrounding fields, causing damage to crops and cultivated lands (Aryal, 2007). While communities living in the immediate vicinity of the Project Area have suffered losses, other interventions to mitigate negative impacts have not achieved the expected benefits, including employment opportunities within the Project Area and social and economic benefits from international tourism and pilgrimage.

There are several initiatives in Lumbini that have attempted to provide responses to the lack of community engagement in natural and cultural heritage conservation activities, some of which have been successful. For example, the Lumbini Crane Conservation Centre and the Lumbini Social Service Foundation have focused on awareness-raising programmes in local schools as well as developing local handicraft projects to strengthen links between local communities within the Lumbini Crane Sanctuary (UNESCO, 2013, p. 152). Other initiatives, notably from IUCN Nepal, UNDP, WWF Nepal, the Department of International Development (DFID), and the Stichting Nederlandse Ontwikkelingsorganisatie Netherlands Development Organization, have focused on the protection and revalorization of the wetlands within and outside the Project Area (Suwal and Bhuju, 2006). WWF Nepal has also supported a local NGO Green Youth Lumbini to promote conservation of sarus cranes and trees. These projects have had positive impacts on the breeding grounds for different bird species and have also contributed to improving waste management and the introduction of renewable and clean energy sources for local communities (UNESCO, 2013, p. 152). Nonetheless, beyond these individual initiatives, there are still long-standing challenges to the development of an integrated approach to the protection of cultural heritage and the environment which involves the communities of Lumbini.

Currently, Lumbini's management does not formally host processes for consultation and conflict resolution between site managers and local stakeholders. Without them, the management of the Project Area remains disconnected from development within the five- by five-mile buffer zones. The Integrated Management Framework, prepared for the World Heritage property in 2013, provides responses to formalize consultation processes within the site's management structure. It has been submitted to the Cabinet of the Government of Nepal for approval, but ratification is still pending. The document redefines the primary objective of site management as the protection of the World Heritage Site, but also refers to the integration of the site in sustainable social and economic development in Lumbini and the wider region. The document introduces new processes to integrate the different stakeholders within the management of the site and the achievement of these sustainability objectives. While the LDT as site manager remains the main responsible authority, with the Department of Archaeology acting as the national authority, the framework has also created associated authorities, including the Lumbini Cultural Municipality (Weise, 2013, p. 10), that could formally join site management within established consultation and conflict resolution processes. The adoption of these recommendations through a formalized consultation process could certainly contribute to addressing some of the long-standing challenges in environmental conservation within the Project Area by providing a platform for resolving conflicts related to the management of natural resources between site managers and local communities and for co-ordinating future projects and activities.

Lumbini's Master Plan and the Environmental Impacts of Pilgrimage and Tourism

Since the conception of the LMP, increasing pressure has been placed on the Western Terai's environment and its local cultural and natural heritage, such as from rapid increases in population, industrial development, infrastructure construction, and over-extraction of natural resources, notably gravel, sand, and soil extraction. New agricultural practices, including the introduction of new crops, agricultural methods, and machinery, are also affecting the surrounding landscape, its traditional livelihood, housing, and the ways of life of communities in Lumbini and its wider region.

Increased air, water, and soil pollution also represents a major impact, with negative consequences on local environment, heritage, and public health. Data collected by the International Centre for Integrated Mountain Development (ICIMOD) air quality monitoring station in Lumbini indicates that the concentration of key air pollutants often exceeds national and international limits, 'which implies significant health risks for the residents and visitors in the region' (Rupakheti *et al.*, 2017, p. 11041). For certain pollutants, the concentration in Lumbini has even regularly exceeded levels in central Kathmandu. Preliminary studies have also indicated that the main sources of air pollution in Lumbini are fossil fuel combustion from transport, industries and residential sources, and biomass burning, including agro-residue burning, waste burning, and regional forest fires (Rupakheti *et al.*, 2017, p. 11048). These studies have traced the origin of the pollutants, suggesting that a large proportion originated from the industrial regions of the Indo-Gangetic plains. However, a significant share could also be attributed to emissions within the Terai, linked to increased agricultural activities, transport, industries, and population densities (Adhikary *et al.*, 2019, p. 204).

Contributing to these environmental issues, waste management is an important challenge in Lumbini that affects both the pilgrim experience and the long-term preservation of the ancient monuments. Visitors' interviews have indicated that waste management is one of the main concerns as it disturbs the spiritual atmosphere of the site (UNESCO, 2013). At the site level, although picnicking is prohibited in the Sacred Garden and additional disposal facilities have been provided, the management and regular cleaning of offerings remain an issue (UNESCO, 2013, pp. 153, 173). The UNESCO/Japanese-Funds-in-Trust project conservation report indicated that the quantity of offerings at the base of monuments and their prolonged contact with organic, plastic, and chemical materials has affected the conservation of these monuments. Moreover, waste management outside the Sacred Garden is also a major challenge in the Project Area, the Municipality, and the wider region, and has been amplified by the rapid increase in visitor numbers.

Rapid industrial development also contributes to the worsening of air, soil, and water quality in the Greater Lumbini Area. In Lumbini, the UNESCO/Japanese-Funds-in-Trust project conservation team has notably identified the deposition of minerals, usually associated with the cement industry, on the surface of the Asoka Pillar which affects its preservation (Meucci, 2013, p. 16). Patterns of industrial development within the wider region have closely followed the main road axis originally developed as part of the LMP regional pilgrimage components, connecting Bhairahawa-Lumbini-Taulihawa, as it has facilitated transport and access to the wider markets (UNESCO, 2013, p. 154).

At the regional level, there is limited coordination between site management and existing legal provisions for the control of local and regional development, although both national legislation, including the Environmental Protection Act, and regional regulations are in place to regulate industrial activities. A broader restricted area was established in 2009, the Lumbini Protected Zone, covering an area of 828 square kilometres, within which carbon-emitting industries are legally banned (UNESCO, 2013, p. 154). However, enforcement of this legislation has been problematic, and uncontrolled growth still represents a major threat to the regional environment and the World Heritage Site. In 2011, IUCN conducted an Environmental Impact Assessment which provided key recommendations and suggested mitigation measures for the reduction of air, water, noise, and soil pollution (IUCN, UNESCO Kathmandu and OCHSPA, 2013). The study recommended shifting the existing carbon emitting industries outside the Lumbini Protected Zone with the provision of compensations and alternative locations for the owners. The latter recommendations have yet to be implemented.

Visitor numbers in Lumbini have been forecasted to increase from 1.2 million in 2017 to 3.4 million by 2025 (Government of Nepal, 2014, Table 8). To support this development, regional infrastructure projects, funded by the Asian Development Bank (ADB), are ongoing and focus on upgrading the domestic airport in Bhairahawa into an international airport and extending the Bhairahawa-Lumbini-Taulihawa road. The airport upgrade is delayed and was still ongoing by the initial completion date, in summer 2019 (ADB, 2018). The roadwork should be completed by 2022 (ADB, 2019, p. 10). These projects have raised concerns from stakeholders. The concerns are partly related to the threat that rapid infrastructure development represents for the preservation of the regional subsurface heritage, notably damage by encroachment and soil extraction for construction purposes. Major concerns have also been voiced regarding the environmental impacts caused by increased traffic, but also indirectly by the expected growth of industries along improved transportation networks and increases in visitor numbers.

Although these two projects are complementary and geared towards similar objectives of facilitating access to the Western Terai and Lumbini to a larger number of visitors, each has been evaluated separately within the ADB's environmental and social impact assessment, monitoring, and evaluation procedures (ADB, 2009). ADB evaluated the environmental impacts of both projects as limited, primarily 'site-specific', and not requiring an in-depth Environmental Impact Assessment (ADB, p. 19). The initial environmental examinations for the airport upgrade, for example, concluded that 'air impact from pollution will be direct, of low significance, local and long-term for short periods only' (ADB, 2013, p. 31). However, the study overlooks the indirect impacts of the airport construction and expected increase in visitor numbers, notably on land traffic, and the cumulative impacts from the major ongoing roadwork and related industrial development along the main corridors. Concerns have also been expressed by other international organizations, notably UNESCO, regarding future landing and take-offs paths which could involve aircraft flying over the Sacred Garden (ISC, 2017, p. 10). At low altitude, planes would be seen and heard from the Sacred Garden, disrupting the tranquillity and sanctity of the site and the visitor experience.

The multiplicity of conflicting reports and uncoordinated interventions by different national, regional, and international agencies represents an important challenge for the management of regional impacts and the development of a holistic approach. As illustrated above, the largest investment currently focuses primarily on transportation and accessibility for increasing visitor numbers, while other key tourism sectors, including accommodation, retail, and the wider visitor offer, have not been incorporated in the regional tourism infrastructure development plans. The repercussions of this are significant not only on the visitor experience but also on stakeholders' capacity to plan and control development and monitor environmental and heritage impacts. The role of the international community, including international institutions and national foreign aid organizations, as key providers of technical and financial assistance, is central, but it does not currently form a united body. Since 2011, UNESCO has increasingly taken a coordinating role with the organization of the annual meeting of UNESCO's International Scientific Committee for Lumbini, bringing together different partners and actors, including key ministries and municipalities, ambassadors from donor countries, and Buddhist organizations, to agree on recommendations for future developments (ISC, 2017). Membership was specifically expanded in 2015 to ensure that associated agencies, such as the Ministry of Urban Development, the Ministry of Local Development, the Lumbini and Kapilavastu municipalities, and ADB were also involved. One of the areas of focus of the Committee's recommendations has been on environmental policies, including the monitoring and control of sources of air, noise, and ground water pollution. The members of the Committee have reiterated the importance of enforcing existing regulations, including the prohibition of new factories within the Lumbini Protected Zone and the relocation of polluting industries outside the Zone as per IUCN's recommendations (ISC, 2017).

It is also clear that there is a gap in the institutional framework to address uncontrolled developments in the region and its environmental and heritage impacts. The Lumbini Integrated Management Framework provides management processes at the local level but is not equipped to integrate the different levels of governance to effectively manage, monitor, and evaluate the sustainability of the wider regional development. To bridge the existing institutional gap, the creation of a regional plan for the Greater Lumbini Area, covering an area of about 85 by 25 kilometres across the three neighbouring districts with rich archaeological heritage – the Kapilbastu, Rupandehi, and Nawalparasi Districts – is currently being investigated to provide the currently lacking regional framework. In its conceptual form, the institutional framework would involve the establishment of a single coordinating body by the Government of Nepal, with the authority and capacity to plan and implement a regional plan in cooperation with international experts and the different associated national and regional agencies, as well as align development objectives with heritage and environmental protection policies. The planning process has been initiated within the third phase of the existing UNESCO/Japanese-Fund-in-Trust project, which is currently conducting detailed surveys of heritage resources in the entire area and investigating key sites in the region to inform the creation of a 'controlled regional development map' (Weise, 2012, p. 1).

Conclusion

The modern development of Lumbini provides an example of how the natural setting and design of the environment around a key pilgrimage site was given a prominent role in defining the religious and cultural experience for pilgrims and visitors. The LMP provided a vision for the site development that reflected Buddhist philosophy and the events in the life of the Buddha through the creation of forests, gardens, and water bodies. Framing connections between textual traditions and the sacred and archaeological remains, the LMP also integrated tourism infrastructure development within the site. In many ways, the design embodies links that have been recently recognized between Buddhism and ecology, particularly the importance of linking ecologically friendly concepts of non-violence and the alleviation of human and animal suffering (Dorje, 2011; Shaw, 2016).

The implementation of the plan has transformed the landscape around Lumbini, with new flora and fauna being harboured, notably endangered species. However, the implementation phase has faced several challenges, including delays in construction work and poor design of the natural setting, especially forested and conservation areas within the new environment. While solutions have since been identified for some of these issues and are currently being implemented, some of these challenges are linked to broader dynamics outside the Project Area, including the relationship between site managers, conservation activities, and local communities, as well as wider regional development interventions.

While the LMP recognized the 'interdependence' between the site development and regional and national plans (KTU, 1978, p. 8), the document is primarily a physical infrastructure planning exercise and has therefore provided limited guidelines to consider the links between infrastructure development within the Project Area and its local economic and social impact as well as its integration within regional development plans. As the current dynamics between site managers, environmental conservation, and local communities have indicated, this gap in the management process has repercussions on the realisation of the LMP and the naturalized environment it envisioned due notably to competing uses by different stakeholders. Several recent programmes at nearby sites aiming to strengthen community engagement in heritage protection offer new avenues to explore for Lumbini management (Lewer *et al.*, 2019; Suwal, 2019). Most notably, Chitwan National Park World Heritage Site has been particularly successful in developing programmes and initiatives towards conflict resolution and local participation in environmental conservation (Suwal, 2019). The buffer zone has played a critical role in this process, with the park management initiating community forest management systems, along with other people-park and grass-cutting programmes, to the benefit of surrounding communities.

Moreover, the LMP restrictions regarding development have become insufficient to protect the site and its surrounding landscape from deteriorating environmental conditions within the Greater Lumbini Area. The changing regional context with the rapid increase in population, industrialization, changing agricultural practices, increasing visitor numbers, and related infrastructure development are putting pressure on the local environment, affecting the visitor experience, heritage conservation, and more broadly affecting the natural and cultural heritage, hydrology, and fauna and flora of the region, with important social consequences for local communities.

An analysis of the LMP implementation since the late 1970s and current dynamics between stakeholders indicates that the lack of integration of the project with local and regional development plans limits the capacity of site managers and responsible authorities to develop effective responses and mitigation measures for developments that affect the site, its natural setting, and cultural heritage. Strengthening the institutional links and integrated management processes with other stakeholders and actors in regional development at local, regional, and national levels, and with the various international agencies involved in the region, will be a determinant factor in achieving the current objectives for Lumbini to act as a catalyst for sustainable development in the Greater Lumbini Area.

References

ADB (2009) Safeguard policy statement. Available at: https://www.adb.org/sites/default/files/institutional-document/32056/safeguard-policy-statement-june2009.pdf (accessed 26 February 2019).

ADB (2013) Proposed Loan for Additional Financing and Administration of Grant for Additional Financing Nepal: South Asia Tourism Infrastructure Development Project. Initial Environmental Examination Report. Available at: https://www.adb.org/sites/default/files/project-document/79356/39399-034-iee-01.pdf (accessed 26 February 2019).

ADB (2018) South Asia Tourism Infrastructure Development Project (Bangladesh, India, and Nepal): Project Data Sheet. Available at: https://www.adb.org/projects/39399-013/main#project-pds (accessed 20 November 2018).

ADB (2019) NEP: SASEC Road Improvement Project. Environmental Monitoring Report: Semi-annual Report (January-June 2018). Available at: https://www.adb.org/sites/default/files/project-documents/48337/48337-002-emr-en_0.pdf (accessed 11 March 2019).

Adhikary, B., Puppala, S.P., Rana, A., Thomas, S. and Panday, A. (2019) A cloud over Lumbini. In: UNESCO. (ed.) *The Lumbini Sacred Garden*. UNESCO, Paris, pp. 203–210.

Allchin, F.R. and Matsushita, K. (1969) *Report for the Lumbini Development Project, Dec 1969*. UNDP, New York.

Aryal, A. (2007) Blue bull (Boselaphus tragocamelus) in Lumbini – a World Heritage Site of Nepal. *Tigerpaper: Regional Quaterly Bulletin on Wildlife and National Parks Management* 34(2), 4–8.

Badiner, A. (1990) *Dharma Gaia: A Harvest of Essays in Buddhism and Ecology*. Parallax Press, Berkeley, California.

Ball, L. (2019) Saving the Sarus: preserving the Lumbini Crane sanctuary. In: UNESCO. (ed.) *The Lumbini Sacred Garden*. UNESCO, Paris, pp. 193–202.

Baral, H.S. (2019) Birds and wildlife of the Lumbini area. In: UNESCO. (ed.) *The Lumbini Sacred Garden*. UNESCO, Paris, pp. 173–192.

Coningham, R.A.E. and Acharya, K.P. (2013) *Identifying, Evaluating and Interpreting the Physical Signature of Lumbini for Presentation, Management and Long-Term Protection: Report of the Third Season of Field Operations*. UNESCO office, Kathmandu, Nepal.

Chapple, C.K. (2017) Buddhist interdependence and the elemental life. In: Hart, J. (ed.) *The Wiley Blackwell Companion to Religion and Ecology*. John Wiley and Sons Ltd, Hoboken, New Jersey, pp. 90–103.

Coningham, R.A.E., Acharya, K.P., Strickland, K.M., Davis,C.E., Manuel, M.J. *et al.* (2013) The earliest Buddhist shrine: excavating the birthplace of the Buddha, Lumbini (Nepal). *Antiquity* 87(338), 1104–1123. DOI: 10.1017/S0003598X00049899.

Dharmodaya Sabha (1956) Report of the Fourth World Buddhist Conference of the World Fellowship of Buddhists. *Fourth World Buddhist Conference of the World Fellowship of Buddhists*, Regional Centre of the World Fellowship of Buddhists, Kathmandu, Nepal, 15-21 November 1956.

Dorje, O.T. (2011) Walking the path of environmental Buddhism through compassion and emptiness. *Conservation Biology* 25(6), 1094–1097. DOI: 10.1111/j.1523-1739.2011.01765.x.

Gopi Sundar, K.S. (2010) Cranes, Cultivators and conservation. *Nature without Borders Symposium Seminar* 613, 34–37.

Government of Nepal (2014) Tourism Promotion Plan for Lumbini and Adjoining Areas towards Making the Buddha's Birth Places & Associated Sites a Regional Tourism Hub (2015-2024 AD). Unpublished report.

ICOMOS (1996) *World Heritage List: Lumbini (Nepal) No* 666 rev. Available at: https://whc.unesco.org/en/list/666/documents/ (accessed 10 November 2018).

ISC (2017) International Scientific Committee UNESCO/Japanese Funds-in-Trust for Strengthening the Conservation and Management of Lumbini, the birthplace of Lord Buddha, Buddha Maya Garden Hotel, Lumbini, Nepal, 17 to 19 February 2017: Recommendations. Available at: http://www.unesco.org/new/fileadmin/MULTIMEDIA/FIELD/Kathmandu/pdf/ISCMeetingRecommendationsFeb2017.pdf (accessed 28 March 2019).

IUCN, UNESCO Kathmandu and OCHSPA (2013) Environmental Impact Assessment of Industrial Development around Lumbini, the Birthplace of the Lord Buddha, World Heritage Property. Available at: https://unesdoc.unesco.org/ark:/48223/pf0000219616 (accessed 10 March 2019).

KTU (1978) *Master Design for the Development of Lumbini, Phase II. Final Report*. UN Archives, New York.

Lewer, N., Lafortune-Bernard, A., Coningham, R.A.E., Acharya, K.P. and Kunwar, R.B. (2019) Community engagement in the greater Lumbini area of Nepal: the Micro-Heritage case study of Dohani. In: Coningham, R.A.E. and Lewer, N. (eds) *Archaeology, Cultural Heritage Protection and Community Engagement in South Asia*. Palgrave Pivot, Basingstoke, UK, pp. 59–74.

Meucci, C. (2013) Conservation of archaeological remains: final report. Report submitted to the strengthening conservation and management of Lumbini, the birthplace of Lord Buddha, world heritage property UNESCO project FIT/536NEP4001 funded by the Japanese Funds-in-Trust for the preservation of the world cultural heritage. Available at: https://unesdoc.unesco.org/ark:/48223/pf0000225760 (accessed 8 March 2019).

Molesworth, K. and Müller-Bölke, U. (2005) The local impact of Under-Realisation of the Imp: a field report. *Contribution to Nepalese Studies* 32(2), 183–211.

Nishimura, Y. (2013) Review of the Kenzo Tange master plan for the sacred garden: final report. UNESCO office, Kathmandu. Available at: https://unesdoc.unesco.org/ark:/48223/pf0000225765 (accessed 8 March 2019).

Pandey, R. (2007) Community multimedia centre Buddha Nagar: Lumbini world heritage site. In: Martin, K., Koirala, D., Pandey, R., Adhikari, S. and Acharya, G.P., *et al.* (eds) *Finding the Local Community Media: Some Stories from Nepal*. ASIARIGHTS Journal 8, 13-20.

Rupakheti, D., Adhikary, B., Praveen, P.S., Rupakheti, M., Kang, S. *et al.* (2017) Pre-monsoon air quality over Lumbini, a world heritage site along the Himalayan foothills. *Atmospheric Chemistry and Physics* 17(18), 11041–11063. DOI: 10.5194/acp-17-11041-2017.

Shaw, J. (2016) Religion, 'nature' and environmental ethics in ancient India: archaeologies of human:non-human suffering and well-being in early Buddhist and Hindu contexts. *World Archaeology* 48(4), 517–543. DOI: 10.1080/00438243.2016.1250671.

Shrestha, M. (1995) Human attitudes and cranes in Nepal's Terai region. In: Schimmel, C. (ed.) *The proceedings of the Third Annual International Crane Symposium: People Water and Wildlife: Human Population Impacts on Cranes, 17-18 March 1994.* National Audubon Society, Boulder, Colorado, pp. 53–55.

Suwal, R.N. (2019) Community engagement in natural heritage conservation stewardship, Nepal. In: Coningham, R.A.E. and Lewer, N. (eds) *Archaeology, Cultural Heritage Protection and Community Engagement in South Asia*. Palgrave Pivot, Basingstoke, UK, pp. 59–74.

Suwal, R.N. and Bhuju, U. (2006) Lumbini: an environmental analysis. In: UNESCO. (ed.) *Lumbini Present Status and Future Challenges*. UNESCO office, Kathmandu, pp. 89–100.

UN (1970) *Lumbini the Birthplace of the Buddha*. United Nations, New York.

UNDP/UNESCO (2013) UNDP/UNESCO Lumbini support project: background information on Lumbini and the greater Lumbini area. Prepared by UNDP/UNESCO Lumbini Support Project, November 2013. Available at: http://unesdoc.unesco.org/images/0022/002252/225236E.pdf (accessed 12 November 2018).

UNESCO (2013) *The Sacred Garden of Lumbini: Perceptions of the Buddha's Birthplace*. UNESCO, Paris.

Watson, M. and Rai, S.K. (2019) Botany and the Buddha: the plants of Lumbini. In: UNESCO. (ed.) *The Lumbini Sacred Garden*. UNESCO, Paris, pp. 151–172.

Weise, K. (2012) Greater Lumbini "Culture for Development" Strategy. Unpublished report.

Weise, K. (2013) Strengthening Conservation and Management of Lumbini, the Birthplace of Lord Buddha, World Heritage Property: Integrated Management Framework. UNESCO office, Kathmandu. Available at: http://unesdoc.unesco.org/images/0022/002257/225755e.pdf (accessed 11 November 2018).

WHC (2004) Decisions Adopted at the 28th Session of the World Heritage Committee (Suzhou, 2004). Available at: https://whc.unesco.org/archive/2004/whc04-28com-26e.pdf (accessed 13 April 2019).

7 Interreligious Dialogue: Trees, Stones, Water, and Interfaith Ritual Experiences in Lebanon

Nour Farra-Haddad*
St-Joseph University, Beirut, Lebanon

Believe in the stone and you will heal (*Amen bil hajar btechfa*)

– A Lebanese proverb

Introduction

Located between Syria and Israel/Palestine, Lebanon has a long history of religious diversity. This is in part because, as a part of the Fertile Crescent and its access to the Mediterranean Sea, Lebanon has been a crossroads for various cultures, ethnicities, and religious groups. Today, hundreds of historical and archaeological sites can be found in Lebanon, reflecting both ancient and modern world history and its religious mosaic. Officially, there are 18 recognized religious denominations. These denominations include four Muslim sects (Shia, Sunni, Alawites, and Ismaili Shia), the Druze (Al Muslimin al Muwahhideen, or 'Muslims, believers in one God'), 12 Christian sects (Maronites, Greek Orthodox, Armenian Orthodox (Gregorians), Armenian Catholics, Syriac Orthodox (Jacobites), Syriac Catholics, Assyrians (Nestorians), Chaldeans, Copts, evangelicals (including Protestant groups such as Baptists and Seventh-day Adventists), Roman Catholics, and Judaism.

While these communities do have a history of conflict throughout Lebanese history, there are many examples of interfaith dialogue and peaceful coexistence in this country. Popular pilgrimages or visits to religious sites or to the tombs of saints have spread beyond their orthodox Christian and Muslim forms and today are an expression of people's piety and their need to create a relationship link between their daily existence, their earthly problems, and the divine. In contrast with the codified religiosity of the mosque and church, believers in many of these faiths have developed a less constraining religiosity, and often join with members of other faiths to worship together. In fact, it is not surprising to see Muslims, Christians, Druze, and Buddhists visiting the same oratories, chapels, monasteries, mosques, maqâms, and mazars looking for *baraka* (i.e. blessings and grace) through praying to the same saints and engaging in the same religious rituals. Each sacred site offers a series of devotional practices to all these faithful, which involve reaching, touching, feeling, and soaking up the *baraka* of the Saint (Farra-Haddad, 2015, 2016, 2017).

This shared worship in sacred space is a good example of inter-religious or interfaith dialogue. Inter-religious dialogue can be defined as various forms of interaction and dialogue, whether formal or informal, between different religious groups (cf. Abu-Nimer *et al.*, 2007; King, 2010). While religious faiths have probably engaged in dialogue among themselves for

*nour@neoslb.com

centuries, it is believed that the 1893 World and the Holocaust served as catalysts for further research in how to different religions interact with each other and what the 'describe[ed] and prescribe[ed] ... proper relationship between religions' should be (Cornille, 2013, p. xii). Inter-religious dialogue, while covering a wide range of interactions between faith traditions, entails that faith leaders and participants 'engage one another from a faith position, and in that the goal of dialogue lies not only in mutual understanding, but also in the pursuit of truth and of personal and religious growth' (Cornille, 2013, p. xii).

Inter-religious dialogue can be divided into different types. For example, Knitter (2013) (see Mitchell, 1993) distinguishes between four types of interreligious dialogue:

- the *dialogue of theology*, which focuses on understanding the beliefs, doctrines, and practices of other faiths
- the *dialogue of spirituality*, which focuses on the experiences derived from these beliefs, doctrines, and practices
- the *dialogue of action*, where people from different faith traditions work together to solve common problems
- the *dialogue of life*, which focuses on how faith traditions interact on a daily basis.

In addition, Eck (1986) distinguishes two dialogues: parliamentary (referring to broad organizations and assemblies, such as the Parliament of the World's Religions, that coordinates interfaith dialogue); and institutional (when individual religious institutions initiate interfaith discussions). As such, interfaith dialogue acts as a 'third space' in which discussions and cooperation regarding ethical and social matters that lead to potential acts of solidarity towards peace can be accomplished (Groff, 2002; Chrabieh Badine, 2009; Zaver, 2013). In each of these cases, there is an assumption that each participating faith tradition believes that there is some sort of common ground that can be found and of the possibility of learning from each other (Cornille, 2013; Knitter, 2013).

Shared rituals and experiences among different faith traditions, part of the interfaith 'dialogue of life', can also be part of the 'dialogue of religious experience' (Chrabieh Badine, 2009), which refers to participating in shared religious rituals and experiences. Here I refer to those shared rituals and experiences as 'natural dialogue', in which I emphasize the spontaneous, naturally occurring interfaith dialogues between devotees of different religious communities that visit the same religious shrines. These interfaith, spatial, and natural encounters are a component of 'living together' (Arabic: *al-'aish al-mushtarak*), where interactions between faith communities are a part of daily life. Gibran Kahlil Gibran (1947), a Lebanese poet and thinker, expresses a strong testimony about the Lebanese experience of "living together" in spite of the war episodes and massacres that have taken place in Lebanon:

> You are my brother and I love you. I love worshipping in your church, kneeling in your temple, and praying in your mosque. You and I and all are the children of one religion, for the various paths of religion are but the fingers of the loving hand of the Supreme Being, extended to all, offering completeness of spirit to all, anxious to receive all.

Religious faiths in Lebanon have a long history of performing devotions at sacred sites – where Muslims visit Christian churches and shrines, and vice versa. These joint pilgrimages revolve around feast days dedicated to saints in both faiths, special religious occasions, or through individual votive visits in which pilgrims make a promise or vow to a saint in return for divine assistance with health problems, infertility, or other personal or societal issues. Small *maqâms* (Islamic shrines built on the site associated with a Muslim saint or religious figure), Christian chapels in rural areas, renown sanctuaries and popular urban shrines are visited by thousands of pilgrims every day. Visits to the places associated with saints (*ziyârât*) are said to be imbued with *baraka* – the divine blessing or miraculous force of grace that is present at that location because of its association with the saint. Through their participation in *ziyârât*, pilgrims hope the saint will listen to their problems and requests. They address the saints as confidants, believing that they understand the difficulties pilgrims present to them. During these pilgrimages, believers meet devotees from other faith traditions who have similar problems. Upon this interaction, the visitors often wish each other success with their petitions by saying: '*ziyârâ Makbouleh*' ('May your visit be accepted and fruitful'). As such, religious pilgrimages in Lebanon are for the most part polyvocal, in that they consist of multiple religious voices raising their cares and concerns to the same saints.

The practicalities of interfaith dialogue are therefore truly experienced at the level of shared pilgrimages, where the construction of a faith-based commonality materializes by being a believer, by sharing similar physical and spiritual needs and interests, and by seeking similar experiences in the same space. In doing so, understanding, harmony, and trust between individuals and communities can be strengthened and lead to lasting goodwill and peace between seemingly disparate groups (see McIntosh *et al.*, 2020). Having based several years of anthropological field research on interfaith pilgrimages to Lebanese *maqâms* and churches, I have not observed many instances of the Durkheimian view that the purpose of pilgrimage sites is to allow particular religious communities to affirm their own unique religious identity and community. Rather I note the existence of *existential* or *spontaneous communities* in the Turnerian sense (Turner, 1995) at these sites, where a sense of *communitas* is present in these interfaith settings. I have often heard some pilgrims say the phrase '*Allah Wahad*' ('There is only one God') and in the same breath say, '*Kull al-qeddisin fiyon al-barakeh*' ('All saints possess *baraka*'). Pilgrims and sanctuary-keepers I have spoken to emphasize that the shrines and the saints are sacred to all and that saints operate miracles for Christians and Muslims without faith distinction. Believers meet one another and perform the same practices without trying to hide or deny their religious identity in any way; there is no pressure on visitors to conform to a particular religious ideology, with the religious identity of pilgrims highly respected.

The purpose of this chapter is to further explore interfaith pilgrimages in Lebanon through an examination of how elements in the natural landscape, such as trees, water, and stones are utilized by pilgrims. Many religious traditions around the world through human history have utilized these natural elements in religious rituals and religious community-building exercises. The chapter furthermore demonstrates that through interfaith devotions using these natural elements, an inter-sectarian conviviality is experienced by pilgrims that leads to a natural dialogue between them, leading in turn to the development of a culture of peace. This natural dialogue of life is especially vital for promoting a sustainable peace in the region.

Trees, Leaves, and Bark: Sources of *Baraka*

Since antiquity, different religions have used trees as a common symbol or representation of several religious motifs. Trees are traditionally the symbol of rebirth, immortality, a sign of fertility, opulence and health (Eliade, 1959). They also symbolize life in perpetual evolution, rising to the sky and evoking the idea of verticality as demonstrated by Leonardo da Vinci's 'Rule of the Trees' (Jacobsmeyer, 2011). Trees represent the cyclical nature of life, as when they lose their leaves during the winter season. Also, trees connect the three levels of the cosmos: the earth (roots), the earth's surface (bark), and the sky (branches). Trees also bring together all the earth's elements: water, earth, air, and fire.

There are several examples of religious devotions and rituals involving trees around the world, including in the Middle East and Lebanon. For example, Arabs of the Jahiliyya era (i.e. pre-Islamic societies in the Middle East) worshipped sacred trees (Masarwah, 2013), such as the *Samoura* tree – a type of gum tree considered to be home to the *jinns* (i.e. intelligent spirits that rank lower than the angels). After the Treaty of Hudaybiyyahm which was signed by Muhammad representing the state of Medina and the Qurashyi tribe of Medina in 628, pilgrims came to worship the tree under which the treaty had been signed (Weber, 1996, p. 37; see Quran 48:18). Dermenghem (1954) also references several religious rituals related to trees in the Maghreb or North African countries. Some of these rituals include placing sick people in the branches of sacred trees, burning certain foliage, and tying bits of cloth to tree branches, etc. Dermenghem also notes several types of sacred trees, such as olive, carob, fig, and oak trees (Dermenghem, 1954).

Some pilgrims are convinced that leaving something at a holy place guarantees the saint's fulfilment of a specific demand. This ritual of leaving a holy souvenir demonstrates the need of pilgrims to leave their presence physically by a saint to ensure that the saint does not forget the vows the pilgrim has made. The type of objects left and the traditions of leaving these objects behind vary according to the places of worship and the faith tradition involved. Worshippers may leave personal belongings with a symbolic link to the vow at the site, such as leaving behind bras for those who are having breastfeeding problems and bibs for infants with illnesses. Pilgrims often leave fragments of clothing or fabric, which are tied to a sacred olive tree in front of Mazar El Saydeh in the village of Saaydeh (Bekaa Valley), the only Muslim religious site dedicated to the Virgin Mary in Lebanon. Around the Maronite Monastery of Saint Anthony the Great (Mar Antonios al-Kabir) in Hadath (a southern suburb of Beirut), women hang plastic bags filled with stones on olive trees, which stones represent their prayers and wishes. In the esplanade of the Shia Maqâm of El Nabi Omran, Qleileh (South Lebanon), who is venerated as the father of the Virgin Mary, a dried tree trunk is now a receptacle for many Muslim prayer beads and even Christian rosaries. And at the Druze Maqâm of Sayyid, Abdallah El Tannoukhi, one of the most prominent Druze historical figures (Firro, 1992), in Abey, Lebanon, site managers provide pilgrims with small ribbons to be tied to a sacred tree as a ritual expression of their devotion to the saint.

Pilgrims also participate in rituals associated with trees in terms of seeking for or preparing a remedy for illness. In some cases, pilgrims swallow sand, oil, or the fruit from a tree at a holy site. Other pilgrims may take natural elements from holy sites home and make a remedy, such as making herbal teas with boiled sacred water or with tree leaves, acorns, or pieces of bark collected from the worship site. For example, at the Druze Maqâm of the prophet Job (Nabi Ayyoub) in Niha or at the Nabi al-Jalil Mazar in Qana (south Lebanon), both shrines are shared by faithful from different sects and reputed to be miraculous. At both these sites visitors take pieces of bark from the sacred tree believed to contain *baraka* and extract chemical compounds from the bark to make 'medicines'. The Nabi al-Jalil is known by the Shia Muslim local community in Cana as the Father of the bride of Cana, and is associated with the first miracle of Jesus Christ. At the Allaoui maqâm of Nabi Berri, where people go to get their warts healed in north Lebanon, Akkar, and along the trail to the Hermitage of SS Peter and Paul in Annaya where the Maronite Saint Charbel Mahhlouf lived, pilgrims frequently collect acorns from the oak trees to make herbal remedies with sacred water.

The circumambulation rite (*ṯaouaf, ṯoufan, ṯatouaf*), which is characteristic of certain places in the context of votive pilgrimages, consists of circling a certain number of times around a worship site, a tomb, a *darih*,[1] a stalagmite or a tree. Multiple examples of this ritual are noted among Christians as well as Muslims. For example, an oak tree sits on the esplanade of the Mar Sassine Maronite Church in Beit Mery. The roots of this tree are known to be miraculous, and pilgrims pass three times under the roots of the tree in the church square as a part of their pilgrimage rites (Farra-Haddad, 2013).

While many pilgrims leave items behind at these sacred sites, they also seek to bring home the *baraka* of the sacred sites in the form of a sacred souvenir. Such souvenirs, which may include a soil sample, water in a flask, a piece of cloth, tree or plant leaves, pieces of bark, or roots, are taken to help pilgrims maintain their relationship with the saint. Some pilgrims will keep these items on an altar at home and continually seek benedictions and graces from them. Other pilgrims will use these material mementoes to create sacred amulets (*dkhireh, hijab, hjab*). Both Christian and Muslim pilgrims create and wear these amulets or place them on their pillow or in their car. Some devotees and locals make and sell amulets at sacred shrines.

In addition to these rituals, there are certain trees that are held as being more holy or sacred than others. The first type is the oak tree. In many faith traditions, the oak tree is synonymous with moral and physical strength. In Lebanon, the oak tree is found on the esplanade of most churches, particularly in rural areas, and is emblematic of the hospitality that can be found at these sites – where the faithful can rest under its bushy foliage, taking refuge from the sun. Pilgrims circumambulate around the oak trees, collect acorns for their *baraka* or tie fabric or thread to branches. There was at one time an oak tree near the hermitage of SS Peter and

Paul (noted above) known for sheltering Saint Charbel under its foliage. This tree no longer exists at the site. According to popular accounts, this is because the faithful took away the leaves, branches – and even the trunk – as relics (Daher, 1965, p. 145).

The olive tree is a tree of great symbolic richness mentioned by faith traditions in this region, in part because it is one of the oldest fruit trees in the world (Lehner and Lehner, 2016). In Greece, the olive tree was dedicated to the goddess of war, Athena, who planted an olive tree on the Acropolis in Athens. Under the Roman Empire an olive branch was used to symbolize peace. This use of the olive branch derived from Jewish and Christian faith traditions, in which a dove brought an olive branch to Noah to signify the end of the Flood and peace between humanity and God. In Islam, the olive tree is considered the holiest tree, in that it symbolizes the axis of the world. Olive trees are also associated with light, with olive oil long being burned in lamps. Olive oil is also used in sacred anointing in all three faith traditions.

Along with the olive tree, the fig tree is one of the trees that symbolizes abundance. However, in some cultural traditions, the fig tree is also considered a negative symbol. For example, the milky appearance of the juice produced by some varieties of fig have a sexual connotation. This is seen in Greek mythology, where the fig was the attribute of Priape, the god of fertility, and Dionysus, the God of vine, wine, and ecstasy. In Roman tradition, Romulus and Remus were fed in the shade of this tree. In Egypt and India, the fig tree was also linked to fertility and fertility rites. The fig tree is also a commonly cited symbol in the Old and New Testaments, such as in Genesis, where Adam and Eve sew fig leaves for modesty purposes; in the Book of Judges (9:10) as part of a parable given to Abimelech, and the Gospels (Matthew 21:18–22; Mark 11:12–14) where Jesus cursed a fig tree. However, in the Bible, the fig tree is above all the symbol of peace, happiness of life, and unity. In Lebanon, fig trees are very popular in rural communities and figs are eaten fresh and dried. In another example, a shrine dedicated to Saydet el Zahra (Our Lady of the Flower) in Afqa was established in the basement of a Roman temple dedicated to the goddess Venus. In this shrine, pilgrims hang pieces of cloth or clothing on the branches of a fig tree as part of their fecundity ritual, which rites were inherited from similar Roman fertility rituals connected with the goddess Venus (Sauma, 2014).

The Arbutus (Madrone) tree is associated with many sanctuaries in Lebanon. Pilgrims collect pieces of a branch, the trunk, or the bark from this tree for its *baraka*. At the Druze Maqâm of the prophet Job (Nabi Ayyoub) in the village of Niha, legends surrounding this tree, associated with the Nabi or prophet, give the tree a sacred and symbolic dimension that attracts many worshippers. In the forest of oaks behind the Maqâm, where visitors organize picnics, the Arbutus tree, with its reddish marbled color makes this tree stand out from the rest. This tree, with its bark color giving the impression of the tree being sick, reminds pilgrims of the prophet Job who was afflicted with leprosy. The tree also is marked by 'sores', or holes, where worms sometimes emerge. In Muslim tradition, this represents parts of the body of Nabi Ayyoub (Job) that were covered with loathsome sores with many ugly looking ulcers on his face and hands that were full of worms. Some pilgrims circumambulate around the sacred tree numerous times while praying, while others strip the bark off the tree for medicinal purposes. An Arbutus tree is also the site of religious rituals in front of the Mazar of Nabi El Jalil in Cana (south Lebanon), with many pilgrims also taking leaves and pieces of bark for various religious rituals.

Water Rituals: From Ablutions to Absorptions

Since antiquity, religious rituals related to the use of water have occurred. Water is thought to heal, rejuvenate youth, ensure life and purify, because it dissolves and eliminates impurities. Water is generally associated with three dominant symbols: the source of life, purification, and regeneration (Chevalier and Gheerbrant, 1982, p. 374; Caulier, 1990, p. 10; see Chapter 2, this volume). Water is also closely associated with fertility and fecundity (Soni, 2015). Water was frequently used in religious rituals by pagan religions and then later by Christians and Muslims in this region. Pilgrims generally take water from holy sites home to perform different types of rituals. For example, in the case of ablutions, holy water is sometimes sprinkled over a woman's belly before sexual intercourse to ensure fertility, and bathing in holy water is done for healing purposes. Drinking holy water also serves many of these same purposes.

Many religious rituals in Lebanon involve the use of water and springs at worship sites that transmit *baraka*. At many sacred sites, pilgrims drink sacred water and take it home for both making remedies and ablutions (washings) (Dermenghem, 1954, pp. 144–145). Water also plays an important role in visits to graves, healings, and exorcisms. During certain ceremonies and festivals (*moussems*), it is common for the host to sprinkle guests with holy water to protect them from demons or evil events. Many women participate in ritual baths in holy water if they are having trouble having children (Chebel, 1995, p. 149). Many Muslims also purchase 'Zamzam water', which is water brought back from the Zamzam well by Hajj pilgrims. This water, while containing impurities (Alfadul and Khan, 2011; Shomar, 2012), can be found in Lebanon and is often poured into small metal bottles called 'Zamzamieh' which are then buried with the dead (Kheir Eldeen, 2004).

Most Christian and Muslim worship sites in Lebanon have a source of sacred water, which water is necessary for the performance of religious rituals. At some sites, such as inside the rock chapel of Mar Boutros wa Boulos (SS Peter and Paul church) in Akoura, holy water miraculously oozes from a rock column, in which pilgrims bathe (Sauma, 1994, pp. 176–177). In many other cave chapels, water oozing from the cave walls is believed to have therapeutic properties, such as in the Al Qatteen Convent- Saydet El Bzaz-El Dorr (Our Lady of the Breast Shrine) in Tannourine (north Lebanon), in the sacred cave of Cana (south Lebanon), and at Saint Marina Grotto in Qalamoun. In particular, women looking to bear children or who have problems lactating come to these sites to drink the holy water coming from the rock. Other sites are renowned for sacred caves or sacred water that comes from springs, including the spring of Sitt Sha'wana near the Druze Maqâm in Ammiq and the spring of Saydet el Zahra (Our Lady of the Flower) shrine in Afqa. The water in the cave of the Mar Jiryes Al Khodr El Batiyeh Oratory (St George's Church) in Sarba is used to bathe children with dwarfism. The prophet Sulayman's Maqâm, located southwest of the village of Younine, is built near a small spring whose water is considered sacred and containing miraculous healing properties.

There are also several sacred wells in Lebanon that are well known for their healing waters and their legends and myths related to saints and miracles. For example, in Smar Jbeil, the miraculous water of the Mar Nohra well is used by pilgrims who have eye diseases, and at the convent of Our Lady of Ras Baalbek (Deir Saydet El Ras) in the village of Ras Baalbek, there is a miraculous well visited by Christians and Muslims. At the Melkite convent of Mar Elias (Prophet Elijah) in Jiita/Hadeth Baalbeck and at the Monastery of Mar Sassine in Batroun, the tradition was to send children down the well on a rope so they could immerse themselves in the holy water. However, this practice has been abandoned, and pilgrims now take the blessing from the top of the well (Kheir Eldeen, 2004). Other holy wells have dried up, such as the well behind the Al Khodr Mosque in Beirut (Qarantina), which was condemned during the war period. Also, at the entrance of the cave of Saydet El Mantara (Our Lady of the waiting Greek Catholic Shrine) in Maghdouche, a well that was renowned for its miraculous water dried up. However, presently the rector of the sanctuary blesses tap water in the church and offers it in small vials to the faithful for the *baraka*, believing strongly that the water contains the *Barak* of the shrine.

Water from fountains in front of or near worship sites is also viewed as having therapeutic virtues. At Maqâm Nabi Najjoum, east of the village of Tarayya, is a natural water basin that never seems to dry up and keeps the same water level all year round, even when there is no precipitation in the winter months. This water is considered holy or blessed by visitors and worshippers of different denominations, who wash with it to take advantage of its virtues. Another famous fountain that attracts devotees is found at the Sanctuary of the Miraculous Medal in Achrafieh, which has ties to the sanctuary of Our Lady of Lourdes in France. Visitors to this sanctuary in Lebanon drink the sacred water and take it home for its *baraka*.

Healing Rituals and Dry Ablutions: Sacred Stones, Caves, and Soil

Like trees and water, the veneration of stones and their use in religious rituals has occurred from pre-historical times. Whether simple stones, rocks, megaliths, columns, rocks transformed into statuettes or statues, benches or other shapes, stones are a symbol of strength, indestructibility, and sustainability. In some cases, depending on their shapes or colors, the virtues of the stones vary.

For the Arabs, during the Jahiliyya era it was believed that the jinns inhabited springs and stones. As such, some sacrifices were made on such stones. The Ka'ba of Mecca is one of those stones, as are the stones that make up the pillars or altars at Mina (*ansab*). Many Arab texts, such as medical, pharmacological, and chemistry books, as well as texts dealing with magic, alchemy, divinations, and other superstitions, mention stones with therapeutic virtues (Chebel, 1995, p. 336). Christian texts also refer to stones as being sacred (Hasluck, 2000, pp. 176–201). Hasluck (2000), in discussing Christian and Muslim uses of sacred stones or rocks, noted that these faith traditions tend to differentiate between 'natural' stones and 'worked' stones, where natural stones have natural virtues, like the Ka'ba stone, and worked stones have been manipulated by human hands (e.g. statues, reliefs). Two other categories of sacred stones Hasluck discusses are stones that have a unique human-made mark, such as the stone that is believed to bear the footprint of Abraham in Mecca, and sacred stones on which inscriptions have been carved. Dermenghem (1954, pp. 122–123, 127–133) also notes different types of uses of sacred stones in religious rituals, including differences between the stones offered by pilgrims at shrines, holy stones that exist near the tombs of saints and are used for 'dry ablutions', and sacred stones that have been manipulated, such as a bench carved out of rock where the sick lie.

The most common example of the use of sacred stones or rocks in religious rituals in Christian and Muslim shrines in Lebanon is sacred stones called *mahdaleh* or simply *al-ḥajar* (the stone). These stones range in size (can take different forms or sizes as simple pebbles, small antique columns or concretions that come from natural cavities, pieces of stalactites-stalagmites) and are considered to have therapeutic properties if rubbed on an injured part of the body. Several Christian and Muslim worship sites are home to sacred healing stones, including Saydet Beshouat (Our Lady of Beshouat); Mar Jirjis (St Georges) in Amshit; Mar Boutros wa Boulos (SS Peter and Paul) in Akoura; Mar Antonios (St Anthony) in Qoshaya; Mar Edna in Ain Ikrine; Mar Doumit in Fidar; Maqâm El Nabi Nouh (Noah) in Karak-Zahleh; Maqâm El Nabi Ayla (Prophet Elijah); and Maqâm El Nabi Youssef (St Joseph) in Haytla. Sometimes, at sites where there are numerous sacred stones, these stones are piled on top of each other by pilgrims to create a small altar, such as is done at Nabi Hedouan Maqâm in Afka and Nabi Najjoum Maqâm in the Bekaa. Other stones are the focus of circumambulation rites, such as at the rock church of Mar Boutros in 'Akoura or around the column of Mar Semaan (St Simeon) church in the Byblos region, where pilgrims circulate around stone columns. Inside the cave by Saydet Hamatoura (Our Lady of Hamatoura) Orthodox monastery in Kousba, women circumambulate around a sacred stalagmite three times (a phallic symbol) and then wrap some of their clothes around the stalagmite. Some of these sacred stones serve specific purposes. For example, the '*Hajar el hebleh*' stone (Stone of the Pregnant Woman) in Baalbek is rubbed on the body of women who desire children.

There have been instances of thieves trying to steal these sacred stones or the stones going missing. At Saydet Beshouat (Our Lady of Beshouat), numerous attempts to steal the sacred stone have led church leaders to lock the stone inside a metal box built into the church's south wall. Because pilgrims could no longer pass the stone over their bodies, they began to access the stone's *baraka* by simply touching the stone, even when caged in a box. Saint George's Church, located at the Beirut gate (Qarantina), once had a sacred stone, but the stone disappeared when the church was converted into the Al-Khodr Mosque. The miraculous sacred stone of the sanctuary of Saydet El Mantara (Our Lady of the Waiting) in Maghdouche as well as the one of El Imam El Ouzai Maqâm also disappeared a few years ago and have never been replaced.

Pilgrims also bring stones and place them at sacred shrines as an ex-voto offering. These stones are sometimes placed on the ground or hung in bags on the walls of natural cavities or on trees, such as in the olive groves facing the Mar Antonios/St Anthony church in Hadath (South Beirut suburbs). In many maqâms, mainly rural shrines visitors lay stones in the *Darih* (a small room that separates a grave of a saint from the surrounding area) as both an offering to the saint and to remind the saint of the visit of the pilgrim and their request, as at the Maqâm El Nabi Youssef (St Joseph) in Haytla.

Related to sacred stones is the use of caves as grottoes or places of worship, in part because of their mystery and the protection they give from the natural elements (see Chapter 2, this volume). In most religious traditions, caves are closely associated with water and rocks and have been surrounded by a sacred aura associated with life and death (Philibert, 1990, p. 15). The cave is the archetype of the original matrix or uterus: a place of birth and rebirth. In the Christian religion, the cave has symbolic importance: Jesus is born in a stable/cave and cults are devoted to different saints in caves, especially to the Virgin Mary, such as at Lourdes Shrine (De La Varende, 1990, pp. 33–40). In Islam, the cave has a strong symbolic importance, for the first vision Prophet Muhammad received was in a cave called Hira near Mecca (Weber, 1996, p. 141). In Lebanon, many natural and human-sculpted rock cavities or formations are used as churches or oratories. There are several examples of these caves as sacred sites, include Sheikh Abdullah's Maqâm in Baalbek, Sheikh Mousafer's cave at Kherbet Anafar, and Sidi Abou Baghracheh in Qab Elias (Kheir Eldeen, 2004). Sauma (1994) also notes that grottos, usually found in small, shallow caves, are also regularly used as shrines, especially for votive rituals. Examples would include the shrine cave of Saydet El Bzaz (Our Lady of the Breast) in Mar Mikhael, the cave sanctuary of Our Lady of Mantara (Our Lady of Waiting) in Maghdouché, the cave of Saydet Hamatoura (Our Lady of Hamatoura) in Kousba, the cave chapel of Mar Abda El Mouchammar (Saint Abdon) in Zekrit and the cave of Mar Antonios (St Anthony) in Qoshaya (Valley of the Saints).

It should also be mentioned here the importance of sacred soil in and around sacred sites that are also used for religious rituals. A universal symbol of fertility, the soil of the earth represents maternal function – it gives and protects life. Soil in and around sacred sites is believed to contain strong *baraka*. At the sanctuary of Saint Rafqa (St Joseph Monastery) in Jrabta, for example, site managers offer a small plastic envelope filled with soil from around the Saint Rafqa tomb to pilgrims as they leave. At the cave of Maqâm of Nabi Yousha' in Miniyeh (North Tripoli, Lebanon), pilgrims are encouraged to take soil from the cave to the *baraka* of the site home. Because of the strong belief in the *baraka* of soil, both Christian and Muslim traditions hold that the soil of some holy sites in the region has medicinal properties, to the point where some pilgrims seek to swallow it.

Natural Elements as Catalysers of Interfaith Experiences

During a *ziyârâ*, several of the rituals described above may overlap at the same site, but at the same time be organized to form a single votive approach. These ritual practices aim at creating an ambiance favourable to the fulfilment of the petitions of pilgrims through accessing the *baraka* of the natural elements found at these sacred sites. This search for graces, benedictions, and *baraka* has long existed in religious and spiritual traditions, with nature being a source where pilgrims can partake of the power of the divine. Indeed, the concept of *baraka* in Lebanese *ziyârât* is central to pilgrimage, with pilgrims seeking physical contact with the shrine and its natural elements (e.g. water, stones, tree leaves, tree bark) and to bring *baraka* home. To maximize the chances of a petition being granted, this *baraka* must be obtained.

Most of the shared religious sanctuaries in Lebanon have a very long history that dates to pre-Christian and Islamic times. In both Christian and Islamic belief in this region, the *baraka* from each faith tradition's sacred sites can be transferred and hold meaning to pilgrims from the other faith. As one 19th-century thinker wrote (Renan, 1997): '[t]he sacred will replace the sacred.' The *baraka* of these sacred sites has been inherited and transmitted to the faithful for many generations, and it is believed that over time the power of the *baraka* at these sites will increase. While some of the religious beliefs and rituals with specific natural symbols have been adapted from paganism and antiquity, water, stones, trees, caves, and soil are readily used in the present to transmit the graces of saints and god to pilgrims of all faith traditions.

This chapter contributes to the recent works on contemporary phenomena of 'interfaith crossing' in the Mediterranean region (Albera and Couroucli, 2009), which region offers several examples regarding the division of sacred space, shared religious practices and sociabilities, and the interaction between the religious actors and the sacred places that are presently utilized by different religious groups. In Lebanon, it is the natural elements of the religious sites and the emphasis Christians and Muslims place on these elements that act as catalysers of interfaith experiences – where Christian and Muslim pilgrims worship together in shared sacred spaces to partake of the *baraka* of these sites in harmony.

Note

[1] The term '*Darih*' refers to the tomb of a saint, but in some cases the tomb is only symbolic and indicates a sacred presence.

References

Abu-Nimer, M., Khoury, A.I. and Welty, E. (2007) *Unity in Diversity: Interfaith Dialogue in the Middle East*. United States Institute of Peace Press, Washington, DC.

Albera, D. and Couroucli, M. (eds) (2009) *Religions Traversées: Lieux Saints Partagés Entre Chrétiens, Musulmans et Juifs en Méditerranée*. Actes Sud, Arles, France.

Alfadul, S.M. and Khan, M.A. (2011) Water quality of bottled water in the Kingdom of Saudi Arabia: a comparative study with Riyadh municipal and Zamzam water. *Journal of Environmental Science and Health, Part A* 46(13), 1519–1528. DOI: 10.1080/10934529.2011.609109.

Caulier, B. (1990) *L'eau et le Sacré*. Beauchesne, Paris.

Chebel, M. (1995) *Dictionnaire des symboles musulmans*. Albin Michel, Paris.

Chevalier, J. and Gheerbrant, A. (1982) *Dictionnaire des Symboles*. Robert Laffont, Paris.

Chrabieh Badine, P. (2009) Islamic Christian Dialogues in Lebanon. In: Raphael, N. (ed.) *Hyphen Islam-Christianity*. Arab Printing Press, Beirut, pp. 646–655.

Cornille, C. (2013) Introduction. In: Cornille, C. (ed.) *The Wiley-Blackwell Companion to Inter-Religious Dialogue*. Wiley-Blackwell, Chichester, UK, pp. xii–xvii.

Daher, P. (1965) *Charbel un Homme Ivre de Dieu*. Al-Arz, Beirut.

De La Varende, G. (1990) Symbolisme de la grotte. *Les Dossiers de L'histoire Mystérieuse* 13, 33–46.

Dermenghem, E. (1954) *Le Culte des Saints dans L'Islam Maghrébin*. Gallimard, Paris.

Eck, D. (1986) What do we mean by dialogue. *Current Dialogue* 11, 5–11.

Eliade, M. (1959) *The Sacred and the Profane: The Nature of Religion*. Harcourt Inc, Orlando, Florida.

Farra-Haddad, N. (2013) Dismantling religious boundaries by sharing the Baraka through pilgrimages in Lebanon. *The Journal of the British Association for the Study of Religions* 15, 54–73.

Farra-Haddad, N. (2015) Pilgrimages toward South Lebanon: Holy places relocating Lebanon as a part of the Holy land. In: Raj, R. and Griffin, K. (eds) *Religious Tourism and Pilgrimage Management: An International Perspective*. CAB International, Wallingford, UK, pp. 279–296.

Farra-Haddad, N. (2016) Shared religious rituals and pilgrimage sites: a movement beyond the Christian-Muslim divide. In: Panagkos, A. (ed.) *Religious Diversity Today: Experiencing Religion in the Contemporary World*. 2. Praeger, Santa Barbara, California, pp. 89–110.

Farra-Haddad, N. (2017) Shared rituals through ziyārāt in Lebanon: a typology of Christian and Muslim practices. In: Weinrich, I. (ed.) *Performing Religion: Actors, Contexts, and Texts: Case Studies on Islam*. Orient-Institut, Beirut, pp. 37–52.

Firro, K.M. (1992) *A History of the Druzes*. E.J. Brill, Leiden, New York.

Groff, L. (2002) Intercultural communication, interreligious dialogue, and peace. *Futures* 34(8), 701–716. DOI: 10.1016/S0016-3287(02)00015-0.
Hasluck, F.W. (2000) *Christianity and Islam Under the Sultans*. Isis Press, Istanbul.
Jacobsmeyer, B. (2011) Uncovering Da Vinci's rule of the trees. Inside Science. Available at: https://www.insidescience.org/news/uncovering-da-vincis-rule-trees (accessed 4 December 2019).
Kahlil Gibran, G. (1947) *Tears and laughter*. The Philosophical Library, New York, p. 77.
Kheir Eldeen, H. (2004) Al Mazarat oual makamat al islamiah fi al Bikaa (Islamic Mazars and Maqâms in the Bekaa Valley). PhD dissertation, Holy Spirit University, Kaslik, Lebanon.
King, S.B. (2010) Interreligious dialogue. In: Meister, C.V. (ed.) *The Oxford Handbook of Religious Diversity*. Oxford University Press, Oxford and New York, pp. 101–114.
Knitter, P.F. (2013) Inter-religious dialogue and social action. In: Cornille, C. (ed.) *The Wiley-Blackwell Companion to Inter-Religious Dialogue*. Wiley-Blackwell, Chichester, UK, pp. 133–148.
Lehner, E. and Lehner, J. (2016) *Folklore and Symbolism of Flowers, Plants and Trees*. Tudor Publishing, New York.
Masarwah, N. (2013) Religious beliefs in pre-Islamic poetry. *Sociological Study* 3(7), 531–542.
McIntosh, I., Fara Haddad, N. and Munro, D. (2020) *Peace Journeys: A New Direction in Religious Tourism and Pilgrimage Research*. Cambridge Scholars Publishing, Newcastle upon Tyne, UK.
Mitchell, D.W. (1993) A revealing dialogue. *The Way Supplement* 78, 42–53.
Philibert, M. (1990) La naissance du symbole. *Les Dossiers de L'histoire Mystérieuse* 13, 9–22.
Renan, E. (1997) *Mission de Phénicie*. Impériale Printing, Paris.
Sauma, V. (1994) *Sur les pas des saints au Liban*. F.M.A., Beirut.
Sauma, V. (2014) *Sur les pas de la Vierge et du Christ au Liban*. Chemaly, Beirut.
Shomar, B. (2012) Zamzam water: concentration of trace elements and other characteristics. *Chemosphere* 86(6), 600–605. DOI: 10.1016/j.chemosphere.2011.10.025.
Soni, N. (2015) Important fertility symbols and rites (northern India). *Journal of Humanities and Social Science* 29(4), 51–53.
Turner, V. (1995) *The Ritual Process: Structure and Anti-Structure*. Aldine de Gruyter, New York.
Weber, E. (1996) *Petit Dictionnaire de Mythologie Arabe et des Croyances Musulmanes*. Entente, Paris.
Zaver, A. (2013) Third spaces of dialogue for multi-faith realities. *Religious Education* 108(1), 88–100. DOI: 10.1080/00344087.2013.747880.

8 Mimicking Mountains: Antoni Gaudí's Sagrada Família and the Mountain of Montserrat

Josep-Maria Garcia-Fuentes*
School of Architecture, Planning and Landscape, Newcastle University, Newcastle upon Tyne, UK

Introduction

It is well known that certain natural places and environments are held as sacred and evoke and foster religious experiences within both individuals and groups of people. These same environments have influenced the construction of temples and built structures of pilgrimage centres and other destinations of religious tourism. The case of Mount Meru and Hindu temples in India are an example of these connections (see e.g. Mabbett, 1983). These natural and built environments have also had a strong impact on the creation and physical attributes of pilgrimage routes throughout history, which routes in turn have shaped the placing and physical structure of the religious centres, urban areas, economic development areas, and transportation routes throughout the world (Timothy, 2013; Olsen *et al.*, 2018).

However, very little has been discussed and analysed regarding the way in which the natural environment informs the manner in which religious buildings are conceptualized and built and how these environmental ideas circulate and intersect with the placement of the diverse political and ideological contestations that usually take place around these unique sites. As such, the purpose of this chapter is to discuss the interconnections between the natural and built environment, religion, spirituality, religious and secular tourism, politics, and ideology through examining how the mountain of Montserrat influenced the design and building of the Basílica de la Sagrada Família (Basilica of the Holy Family). The Basilica is both an important centre of worship and a religious tourism site, receiving over 3.2 million visitors in 2014 (Marine-Roig, 2015). Known worldwide as a Catholic expiatory temple, it was originally proposed in 1881 by philanthropist and bookseller Josep Maria Bocabella. Its construction began in 1882 under the direction of architect Francisco de Paula Villar y Lozano, whose architectural vision of the Basilica was a gothic revival project in line with the architectural styles of the time. Lozano was only able to oversee the building's foundations and the crypt until disagreements between him and Bocabella led to Catalan architect Antoni Gaudí overseeing the completion of the building project. Unlike Lozano, Gaudí had a very different vision for the Basilica, wanting to link the building's architecture to both the political and ideological campaigns active in Barcelona and Catalonia at the time, which, indeed, were related to wider ideological competitions on the modernization of the state in Spain and Catalonia.

*josep.garciafuentes@newcastle.ac.uk

These were intense cultural, political, and ideological debates across multiple competing social groups, including Republican and Liberal supporters of the Catalan nation and other groups aligned with the wider Spanish positions in which the symbolic shaping of the city of Barcelona and the mountain Montserrat served as a stage background.[1]

Gaudí and his project for the Sagrada Família aligned with the more orthodox Catalan Catholic groups, who claimed the essential connection between the origins of Catalonia as a nation to the imposition of the Catholic religion in its land back in the Middle Ages. This was, of course disputed, since the ongoing project of the Sagrada Família and Montserrat at the time when Gaudí took over the project had multiple competing ideological meanings linked to the struggle between the various political groups who sought to capitalize its symbolism to suit their own interests. To chart the history of the integration of the ideological and political ideas behind the motivations and design of the Basílica de la Sagrada Família during the late 19th and early 20th centuries, this chapter examines the use of Montserrat's mountain as an ideological tool in Spain, Catalonia, and the Americas with the aim to trace the origin of the intricate and multiple ideological and religious symbolism of the mountain that was mobilized by Gaudí in his project for the Sagrada Família. Then, the chapter analyses Gaudí's architecture and its attempt to interpret architecturally the geological structure of Montserrat with the aim to make a bold statement on the religious symbolism of the Catalan mountain within the competing struggle between the different ideological groups that sought to capitalize both Montserrat's and the Sagrada Família's symbolism to suit their own political purposes. Ultimately the goal of this chapter is to discuss the entangled relations between religion, nature, and ideology, and to demonstrate the role that architecture and architectural interpretation of nature play within the mediation of these connections.

The Ideological and Religious Shaping of the Mountain Montserrat Over the 19th Century

Antoni Gaudí's Basílica de la Sagrada Família is deeply connected with the mountain of Montserrat and to the ideological struggle around it as one, if not the most, important national symbol for Catalonia and Spain that was created over the course of the 19th century and than transformed the purely religious symbolism of the mountain of Montserrat into a more complex symbolism able to refer to multiple ideological and religious ideas.

This process was unleashed in the 1850s, when the poet, writer, journalist, historian, liberal politician, and freemason Víctor Balaguer shaped Montserrat as a national symbol for his project to create a new federal nation that would include the peoples of Spain and Portugal. This was a national project aligned with Víctor Hugo's vision of a 'pan-Latin' federation and with the contemporary Italian political reunification process (Mas, 1851; Cuccu, 2004; Pi de Cabanyes, 2004). While travelling around Europe, Balaguer reflected widely on how the built heritage of different countries contained the memories and traditions of nations, albeit memories and traditions that were 'invented by poetry and conserved by credulity' (Balaguer, 1852, pp. 5–6). These ideas on monuments and traditions led Balaguer to attempt to develop a system of symbols based on the reinvention of the former religious monasteries spread across the Spanish nation because if, as Balaguer argued, 'the Orient is a palace, France is a castle, and Italy is a garden, then Spain is a cloister' (Balaguer, 1851, p. 5).

In his book *The Monks and Their Monasteries*, Balaguer (1851) detailed the history, myths, stories and tales of the Spanish monasteries together with their romantic descriptions and attempted to further reconceptualize their religious and non-religious meaning into his broader political vision of Catalonia and Spain. This incorporation involved a reinvention of all old ruined Spanish monasteries that were affected by a series of Ecclesiastical Confiscations (including the famous *Desamortización Eclesiástica de Mendizábal* from 1835) that led to the State becoming the owner of monastic properties in Spain in order to profit from the sale of monastic lands and to fuel a modern liberal economic market (Tomás y Valiente, 1985).[2] In his next book, entitled *Cuatro Perlas de un Collar* (Balaguer, 1853) Balaguer focused on the great Catalan monasteries of Ripoll, Sant Cugat, Santa Maria de Poblet, and Santes Creus, tying their stories and architecture to his political ambitions, in essence trying to make his political arguments more widely accepted among the diverse social and cultural groups of Spain through the combination of religious, historical, and mythical tales, and romantic descriptions of landscapes and architectures of the former religious monasteries. In this way, old monasteries were reinvented by Balaguer and turned into a flexible monument with diverse – and even opposing, meanings, including religious and non-religious understandings. Balaguer, thus, managed to transform old monasteries into a modern symbol that could potentially be shared by the whole national population regardless of their ideology and religious beliefs or faith.

Of all the monasteries, Balaguer focused on Montserrat, the one that he felt was the most important. In fact, Balaguer obsessed over Montserrat to the point that he adopted the literary name of 'The Troubadour of Montserrat', in conscious ambiguity after the mountain and the famous Madonna statue venerated in the monastery. Indeed, the reason for Balaguer's strong interest in Montserrat over other monasteries was a consequence of the unique mountain where the monastery was located, which was the subject of several overlapping sociopolitical readings. One of these readings that affected Balaguer the most came from the German brothers Alexander and Wilhelm von Humboldt, who, while visiting the mountain a decade before the Battle of Montserrat 1811, were impressed by the unique nature of the mountain and its strange geological formations, which Alexander described as a 'big city in ruins'. Rather than focus their attention on the monastery or its Madonna statue, they were rather struck by the happiness of the hermits living in and around the monastery. This happiness, which seemed to come from the interactions between the hermits and their natural surroundings, made Wilhelm argue that the mountain itself produced a strong relationship between the human spirit and the physical landscape that led to spiritual introspection (Humboldt, 1803, 1951, p. 133). Wilhelm discussed with Goethe his experience of the mountain, and their exchange of letters led to a new reading of Montserrat as a site of spirituality as well as one of religiosity (Humboldt, 1803, 1951; Unamuno and Gárate, 1951).

This reading of Montserrat as a spiritual location led Balaguer to define the mountain in a spiritual but not-necessarily-religious manner – an approach that was viewed unfavourably by the Catalan Catholic Church. Balaguer, however, understood this spiritual approach to Montserrat as an opportunity to forge a unique national symbol that could be a reference to other non-religious or anticlerical groups in addition to religious groups, and therefore able to amalgamate ideological differences. Furthermore, Balaguer believed that this reading could make Montserrat an international symbol thanks to this spiritual interpretation of the mountain – a 'symbol for all' (Garcia-Fuentes, 2013, pp. 40–51). Therefore, Balaguer used Montserrat as a constant symbol during the deployment of his political and cultural projects, particularly in his books and newspaper articles. Through these media, Balaguer managed to spread and popularize his ideological project, and in this way he complemented the use of his strong political connections to foster reconstruction projects and to bolster widely his ambitious national construction ideas. The culminating moment of these activities arrived in 1871 when Balaguer became part of the delegation of MPs who travelled to Italy to offer the Spanish Crown to Amadeus of Savoy, the son of the Italian king who was leading the unification of the Italian Peninsula. The parallel between the Italian project and the Iberian one supported by Balaguer is obvious, since the Italian dynasty embodied all the hopes of materializing the Iberian state, mirroring Italian unification.

Through these entwined political and cultural activities, Balaguer not only reshaped the image of the monasteries and the mountain of Montserrat, but he also contributed to change their materiality and architecture. He encouraged the idealized reconstruction of the ruins of certain monasteries by national elites and by the Spanish Royal Family – including an ambitious project to reshape the shrine of Montserrat that was presented to the new King Amadeus I in 1871. Indeed, through the promotion of these architectural works of construction and the restoration of old, ruined monasteries, Balaguer also tried to replace the social taste for classic architectural models with the romantic value of gothic architecture, which architecture mimicked Mount Montserrat. It is therefore not by chance that he photographed himself, together with other poets from Catalonia and the south of France, during an 1868 excursion, in front of the ruined gothic cloister of Montserrat. Balaguer also sought to rebuild Montserrat monastery using gothic architecture (Garcia-Fuentes, 2012, pp. 97–118).

In addition to the mountain itself, Balaguer also wrote extensively about Montserrat's landscape in a passionate style. In writing about his view from a spot called 'The Monks' Balcony', Balaguer wrote:

> [The statues on the balcony] remain there, silent and still, condemned to contemplate forever the magnificent spectacle stretching out before them ... In front of me, all visible at one glance, was Sant Llorenç, the mount of the mysterious Simanya cave; while farther off stood Montseny, so poetically described by Aribau and Rubió ... and marking off this majestic panorama were the towering Pyrenees mountains (Balaguer, quoted by Cornet i Mas, 1863, pp. 282–284).

This text by Balaguer echoed Wilhelm Humboldt's view of the aesthetics of Mount Montserrat, where he wrote: 'if [I] were to live the life of a hermit here, I would delight in learning to distinguish one peak from another and to give them all names. You say hello to them at dawn and say goodbye to them at dusk' (Humboldt, 1803, 1951, p. 133).

The popularization of the unique aesthetic landscape of the Catalan mountain (Fig. 8.1) by Balaguer influenced the first printed postcards of Montserrat that were published in the early 1880s, which included views from 'The Monks' Balcony' and similar landscapes of the mountain. This progressive shift in the image of Montserrat toward new understandings of the mountain beyond the former strictly exclusive Catholic symbolism can further be traced to the publishing of regional guidebooks in the last half of the nineteenth century, which guidebooks showed the expansive viewscapes of the mountain, whereas earlier guidebooks focused more on the monastery or Madonna statue in isolation (Garcia-Fuentes, 2012, pp. 105–110;.

Furthermore, Balaguer's ideological and aesthetic project successfully popularizing both the mountain and the Madonna in Spain eventually also reached the colonies still under Spanish control, including Cuba, Puerto Rico, and the Philippines. In 1871, during the

Fig. 8.1. A late 19th-century panoramic postcard of Montserrat's unique landscape (Josep-Maria Garcia-Fuentes, used by permission).

popularity peak of Balaguer's project, when the Crown of Spain was offered to the Amadeus of Savoy, a group of Catalan settlers living in Matanzas, Cuba, decided to raise a hermitage devoted to Montserrat in one of the hills on the outskirts of the city overlooking the sea. The building of the church within the hermitage was directly influenced by the writings of Balaguer and his descriptions of the mountain and the Montserrat monastery. In particular, the church architects were heavily influenced by Balaguer's description of the Monks' Balcony, which, as noted above, was reproduced in contemporary guidebooks of Montserrat (see e.g. Cornet i Mas, 1863, pp. 282–284). They organized the exterior spaces in front of the hermitage as a balcony facing towards the landscape, and placed a series of statues in the same position as those on the Monks' Balcony in Montserrat.

Montserrat's balcony was therefore replicated in the construction of the new hermitage devoted to Montserrat in the hilly outskirts of Cuba. In the interior space of the hermitage a Romantic altarpiece representing the unique landscape of the mountain of Montserrat was built, reinforcing a growing awareness of the natural assets of Montserrat in close connection to the entangled multiple modern political, ideological, and religious symbolism around the mountain that was shaped by Balaguer to suit his ambitious project.

The Popularization of the Mountain and Competing Ideological Projects

Víctor Balaguer's project meant the shaping of a complex symbolism around Montserrat, both around the mountain and its Madonna, that moved beyond and over the purely religious imagery that dominated the popular and elitist imagination associated with it since medieval times. The previous understanding of Montserrat before the interest in its mountain by the German romantics and by Balaguer was grounded in the cult to the Madonna that had been venerated there since medieval times. This cult was spread worldwide throughout the Spanish colonies in later centuries, in particular in the Americas – following the performance of the first Catholic Mass on the continent in 1493 by Bernat Boïl, one of the hermit priests of Montserrat, who accompanied Christopher Columbus as a representative of the Catholic Church during the explorer's Second Voyage to the New World (Albardaner i Llorens, 2000). The cult of the Madonna of Montserrat spread quickly to all the Spanish colonies, and became particularly powerful during the different independence movements that turned the colonies into their own countries. This was the case in Argentina, Chile, Colombia, Cuba, Ecuador, Nicaragua, El Salvador, Guatemala, Mexico, Peru, and even in Brazil, where churches and shrines had been built under the advocacy of the Madonna of Montserrat. These churches and shrines were in most cases built within unique natural settings that were similar to the sedimentary geological formations of the original mountain of Montserrat. In addition, particularly during the 19th century, many mountains in Latin American countries that were deemed similar to the skyline of Montserrat were dedicated to the Madonna as a part of the use of natural landscapes in the development of national identities. As such, the development of these identities and communities was directly associated to both Montserrat and its Madonna and therefore clearly Catholic and religious in nature.

Humboldt's and Balaguer's revolutionary understandings of Montserrat also took shape in Spain following the collapse of Balaguer's ideological project in 1873, when King Amadeus I abdicated his throne.

Indeed, this abdication led to an upsurge in competing nationalist movements which, for the most part, were very secular in character and sought to use the recent symbolic capital growth around the mountain of Montserrat for their own political gain. In this context, in the late 1870s, a group of clerics with political ambitions created the *El Grup de Vic* or the Vic Group (after the town of Vic near Barcelona where the group was formed; see Garcia-Fuentes, 2018) which sought to counter the secularism of the burgeoning Catalan nationalist movement and define a new Catholic Catalan nationalism through the appropriation of Balaguer's cultural and symbolic universe forged around ruined monasteries. For this group of clerics, and indeed the Catalan Catholic Church, Catalan nationalism was inseparable from Catholicism, and as such *El Grup de Vic* insisted that the idea of Catalonia as a nation should be directly linked to the building of the first monasteries in the Catalan territory (Garcia-Fuentes, 2018).

Furthermore, in order to dominate the symbolism of the famous mountain, in the 1880s the Catalan Catholic Church organized a series of 'patriotic/religious' campaigns to solidify their religious/political ambitions through shaping both the mountain and the monastery on Montserrat, along with other monasteries like Ripoll, as key symbols in their political and ideological conception of the nation/state (Junyent i Rafart, 1990). In addition, the Group organized two religious festivals at Montserrat to tie together the monastery, the Catalan Catholic Church, and the Catalan landscape with the Catalan nationalist political movement. For the Vic Group, the monastery and shrine of Montserrat were inseparable from the mountain, and the mountain itself was viewed as sacred, as God's creation.

These overt attempts by the Catalonia Catholic Church to incorporate Montserrat and its monastery into its campaigns to sway popular opinion regarding Spanish and Catalonian nationalism were opposed by several liberal nationalist movements, who denounced this appropriation by the Church of something that belonged 'to all' (Almirall, 1880, p. 329) and also had become a symbol of these same nationalist movements. In this context, for example, one businessman group took advantage of the juridical situation of the mountain (that was still expropriated and belonged to the State) and planned to build a rack railway that would run to the top of the mountain where would lie a newly-built scientific observatory called the 'Temple of Science' (Garcia-Fuentes, 2016, pp. 68–70). The rack railway project was begun in 1881 and was completed in 1892. In turn, the Catalan Catholic Church continued to emphasize the sacred character of the mountain through continued foot pilgrimages up to the monastery and other hermitages on the mountain, which were seen as differentiating true pilgrims from tourists (Roma i Casanova, 2004; Garcia-Fuentes, 2012, pp. 219–252; Garcia-Fuentes, 2013, 2016). In addition, it was decided to restore older hermitages on Montserrat. In doing so, they used Romanesque architecture to distinguish their religious and national campaigns from more liberal groups that used rather Gothic architecture. Furthermore, in addition to the restoration of ruined hermitages, the Catalan Catholic Church decided to promote the construction of a series of monuments along the paths that led up the mountain – many of which were created according to picturesque theories and techniques (Garcia-Fuentes, 2010) of organization in time and space as in relation to the walk along these paths; to, again, demonstrate the mountain's sacred character.

During this time the building of miniature replicas of the mountain of Montserrat also became popular. Many of these replicas were created to beautify the bourgeois picturesque and romantic gardens of wealthy Catalan families, as well as to display the values these families saw in the mountain according to their affinity with the major ideological groups mentioned above. Similar replicas were placed in the gardens of semi-private buildings, such as hotels, sanatoriums, hospices, and asylums. Indeed, these replicas were both fashion statements and physical manifestations of the intense ideological struggles that were taking place around the mountain of Montserrat at the time. These miniature reproductions were made in Spain, mostly in Catalonia, but also overseas. For example, in 1884, the Catalan community in La Habana, Cuba, decided to build an artificial replica of Montserrat to use for organized Catholic pilgrimages (Fig. 8.2). The decision to build this replica of the mountain was based on the fact that it was easier and cheaper to build a small Montserrat in Cuba than to organize a popular pilgrimage to the real mountain in the Iberian Peninsula.

At the same time, non-religious replicas of the mountain of Montserrat were also built. For example, one of these was built in Ciutadella Park in the city of Barcelona, following the success of the panorama of Montserrat raised next to the park for the World Exhibition of 1888. This occurred during a moment of key symbolic reinvention and construction of the whole city of Barcelona (Garcia-Fuentes, 2016, pp. 63–74). Like the panorama from 1888, the Montserrat miniature in the main public park of Barcelona recalled the Romantic experiences of different viewpoints to contemplate landscape without reference to the religious nature of the mountain. The reproduction in Ciutadella Park quickly became a popular tourist attraction, as revealed by a series of little paintings by Picasso of the mountain when he relocated to Barcelona in 1895. Because of the non-religious resonances of

Fig. 8.2. An illustration of the miniature replica of Montserrat raised by Catalans in La Habana for celebration rituals as published on 22 May 1884 in the magazine *La Ilustración Española y Americana* (year XXVIII, number XIX), p. 316 (Josep-Maria Garcia-Fuentes, used by permission).

this replica, the Catalan Catholic Church denounced vehemently such improper use of the symbolism of Montserrat. As such, these examples of artificial replicas of Montserrat, both in Catalonia and Cuba, not only occurred as a part of the struggle over who was able to take advantage of Montserrat's symbolic capital, but also as an example of the Catalan Catholic Church's attempt to counterbalance the competing explanations of the mountain in the light of secularized ideological reasoning and the natural sciences.

Montserrat, the Sagrada Família and Gaudí's Architecture

It is within this context of religious and non-religious popularization of the reproductions of the mountain of Montserrat that the Catalan Catholic Church, thanks to the architect Antoni Gaudí, converted Montserrat's mountain into an architectural model for their main churches. Gaudí's buildings, such as the Sagrada Família and the Church of Colònia Güell (an unfinished church in the manufacturing suburb in Santa Coloma de Cervelló near Barcelona; see Lahuerta, 1999, 2016), among others, were understood as architectural reproductions of the mountain, incorporating viewpoints and exterior stairs meant for the contemplation of the views and the city (Garcia-Fuentes, 2012, pp. 219–252). These buildings were therefore related strongly to the Catholic religion, to the land and city landscape of Catalonia, and to the cultural, social, and political ambitions of the Catalan Catholic Church.

These ideas are also revealed in the case of Montserrat by, and represented in, the Monumental Holy Rosary of Montserrat along the trail connecting the monastery to the cave where the Madonna was supposedly discovered in the 1880s (Fig. 8.3). The whole intervention was conceived as a whole in 1896 and it was completed in 1816. It comprised 15 monuments each designed by a different architect, including Josep Puig i Cadafalch and Antoni Gaudí, and sculptors like Josep Llimona, Dionís Renart, and the Vallmitjana brothers (Lahuerta, 1999, pp. 254–256). Gaudí was commissioned for the design of the First Mystery of the Rosary – a depiction of the Holy Sepulchre with a statue of Christ Risen – sculpted by Josep Llimona – and the statues of the Three Marys – sculpted by Dionís Renart. All of them were placed within a large cave within the mountain that was carved following the direction of the architect (Lahuerta, 1990, pp. 254–288).

Gaudí's other architectural creations, whether they be churches or residential buildings, were also based on these religious/political debates and ideological struggles of the

Fig. 8.3. A late 19th-century postcard of Gaudí's monument devoted to the Mystery of Resurrection, in Montserrat, which draws a clear parallel between religious and national rebirth by placing the Resurrected Christ and the Catalan flag next to one another (Josep-Maria Garcia-Fuentes, used by permission).

time (Garcia-Fuentes, 2018, pp. 91–108). But it was the Sagrada Família project that became his pinnacle, up to the point that it consumed all his time and devotion from 1883 until his death in 1926. When Gaudí took on the challenge to build the Sagrada Família, he decided to change radically the project initiated by the architect Villar y Lozano despite the foundation and crypt of the basilica already having been built. Gaudí decided to change the gothic architectural design of the existing project, which was associated with ideologies distant to the Catalan Catholic Church, and to develop a new scheme based on his architectural interpretation of the mountain of Montserrat. In doing so, he made a clear statement that the mountain should be interpreted in exclusively sacred Catholic terms. This is, he understood the mountain as sacred, or in other words, the mountain as the temple. Gaudí drew this inspiration in part from Alexander von Humboldt, who first described the mountain of Montserrat as 'the ruins of a giant rocky city' in 1800 (Humboldt, 1951). Beyond Montserrat, the use of the mountain as an architectural model had a more recent antecedent in Eugène Emmanuel Viollet-le-Duc's sketches of mountain structures, in particular the Alps.[3] Viollet-le-Duc's sketching project was well known at that time within the School of Architecture of Barcelona, when Viollet-le-Duc's *Dictionnaire Raisonné de L'architecture Française du XIe au XVIe Siècle*[4] served as an essential text of architectural education until well into the 20th century. Indeed, both Viollet-le-Duc and Gaudì's architectural inspiration are case studies in connecting architecture, geology, biology, and the natural sciences (Bergdoll, 2007). Gaudì's inspiration also came from the fact that the symbolic use of architecture was a popular topic at the School of Architecture in Barcelona during this time, particularly in the context of the symbolic meanings of Indian and Asian architecture, including, for example, references to Mount Meru in Hindu temple architecture. These discussions were further linked to deliberations regarding how people experienced architecture (Garcia-Fuentes, 2012, pp. 232–252). The religious character of nature in Gaudí were ideas fuelled by the diffusion of Ruskin's thoughts within the contemporary Catalan intellectual circles.[5]

Furthermore, Gaudí planned the construction of the Sagrada Família not as a traditional cathedral with a layered architecture, but rather as a set of towers that were raised up individually from the bottom to the top. In this way, each tower and all the rich details in it would detail particular and complex religious and ideological symbols. Moreover, the whole set of towers could be conceptualized as a symbol of the unique morphological topography of the mountain of Montserrat, and this connection became evident as the new towers and facade of the Sagrada Família evoked the distinctive skyline of Montserrat, making it easier for onlookers to see the connection between the natural and architectural mountains of Montserrat. This use of the mountain of Montserrat itself in the design of the Sagrada Família fuelled the construction of other religious buildings by Gaudí and some of his disciples, such as Josep Maria Jujol, which underlined the sacredness of the mountain as a symbol over other liberal and scientific interpretations and appropriations of the mountain. In this way, these religious buildings became manifestos of the principal ideology in the Catalan Catholic Church, wherein the sacredness of the mountain of Montserrat, again, should be seen primarily as a Catholic symbol over any other picturesque, touristic, or political interpretations and appropriations.

However, these campaigns led by the Catalan Catholic Church to appropriate the symbolic capital of the mountain of Montserrat were highly criticized by other political and ideological groups who themselves attempted to provide certain representations or interpretations of the mountain, as noted above in the construction of replica mountains in Catalonia and beyond.

The tensions around Gaudí and the Sagrada Família, as related to Montserrat, increased exponentially their symbolic capital, making them appealing to multiple groups who tried to appropriate their symbolism into their own goals. This was the case, for example, of the passionate campaigns by the poet Joan Maragall and the architect and art critic Josep Pijoan, who published a series of articles on Gaudí's Sagrada Família between 1905 and 1907. These articles were planned to allow the *Lliga Catalanista* – a Catalan right-wing nationalist party – to appropriate the basilica's and Gaudí's symbolism for its own purposes. While the *Lliga Catalanista* viewed Gaudí as an outsider and his architectural project as unappealing, because of the popular approval of the Sagrada Família project, the group sought to expand the basilica's symbolism beyond that promoted by the Catalan Catholic Church so that the mountain and the Sagrada Família would become symbols for all Catalans, whether they wanted it or not, irrespective of their own individual ideology. For this reason, the articles by Maragall (1900, 1905a, 1905b, 1907) and Pijoan (1905, 1906; Roma i Casanovas, 2002) considered Gaudí 'the genius of Catalonia', and also suggested that the Sagrada Família should be considered 'our work' – 'the work of all Catalans', who must collaborate 'efficiently and without hesitation' to ensure the financing and the completion of the project (Roher, 1999). This provoked a tense contestation in the media by the committee running the construction of the Sagrada Família, who saw this attempt by the Catalan Catholic Church, and later that of the *Lliga Catalanista*, as an attack on what according to their ideas belonged to all Catalans and all Spanish without any exclusion or precision (Garcia-Fuentes, 2019, pp. 97–102).

Conclusion: The Myth of Gaudí

Today, the Sagrada Família and Gaudí's architecture does not always exactly resemble Gaudí's original ideological take on the basilica. Indeed, modern interpretations of the building and its architect are the result of a long process of mythification, in which different architects and intellectuals have negotiated their own interpretations to suit their own personal projects or ambitions, which in turn inform present-day touristic narratives.[6] Part of these reinterpretations of Gaudí and his work seems to stem from the strong architectural contrasts between the parts of the building completed by Gaudí during his lifetime and construction work done after his death (Garcia-Fuentes, 2015a, pp. 64–78). This mythification is in many ways a unique example of how heritage, architecture, and urban space are always the changing outcome of cultural and ideological struggles, which constantly change and never seem to end. Today, most tourists visit the Sagrada Família because of its cultural, artistic, and architectural values rather than for religious reasons (Marine-Roig, 2015, pp. 25–37). Indeed, the Sagrada Família is a representative case study of the difficulty of tracing the limits between religious tourism and secular pilgrimages, as well as the entangled edges between the sites of heritage and the sites of pilgrimage and between religious tourism and the natural, historical, and architectural environment (Di Giovine and Garcia-Fuentes, 2016, pp. 1–23). The Sagrada Família as discussed here in its more important nuances reveals the entangled relations between religion, nature, and ideology, and demonstrates the role that architecture and architectural interpretation of nature play within the mediation of these connections.

Notes

[1] For more information regarding the symbolic shaping of Barcelona during the 19th and 20th centuries, see Guàrdia *et al.* (1988) Garcia-Fuentes and Garriga (2012).
[2] As a result of this law, many monasteries, like Montserrat, were vacated by monks and subsequently fell into ruin due to the State's inability to maintain these buildings (Garcia-Fuentes, 2012, pp. 39–60).
[3] For a wider contextualization of the architectural interpretation of mountains and nature in the search for a modern architecture since the late 18th century, see Garcia-Fuentes (2015b).
[4] Dictionary of French Architecture from the 11th to 16th Century (1854–1868).
[5] See for example Montoliu (1903), which is the first volume edited in Catalan on a translation of selected texts by Ruskin. It was preceded by the publication in Catalonia of numerous references by the British author relating nature, art and religiosity.
[6] See for example the works by Dalí (1934), Cirlot (1950), Zevi (1950), Collins (1960) or Torii (1983).

References

Albardaner i Llorens, F. (2000) John Cabot and Christopher Columbus revisited. *The Northern Mariner* 10(2), 91–102.

Almirall, V. (1880) Lo Milenari de Montserrat. I. antecedents Y Preparatius. *Diari Catal*, 329.

Balaguer, V. (1851) *Los Frailes y sus Conventos. Su Historia, su Sescripción, sus Tradiciones, sus Costumbres, su Importancia*. Librería Española, Barcelona.

Balaguer, V. (1852) *Montserrat. Recuerdos Tradicionales e Históricos de este Santuario y Montaña*. A. Brusi, Barcelona.

Balaguer, V. (1853) *Cuatro Perlas de un Collar. Historia Tradicional y Artística de todos los Celebres Monasterios Catalanes: Santa María de Ripoll, Santa María de Poblet, Santas Cruces y San Cucufate del Vallés*. Libería Española, Barcelona.

Bergdoll, B. (2007) Of crystals, cells, and strata: natural history and debates on the form of a new architecture in the nineteenth century. *Architectural History* 50, 1–29. DOI: 10.1017/S0066622X00002860.

Cirlot, J.E. (1950) *El Arte de Gaudí*. Ediciones Omega, Barcelona.

Collins, G.R. (1960) *Antonio Gaudí*. George Braziller, New York.

Cornet i Mas, G. (1863) *Tres Días en Montserrat. Guía Histórico-Descriptiva de todo Cuanto Contiene y Encierra esta Montaña*. Librería del Plus Ultra, Barcelona.

Cuccu, M. (2004) Balaguer i Itàlia (1848-1901). In: Comas, M. (ed.) *Víctor Balaguer i el seu temps*. Publicacions de l'Abadia de Montserrat, Barcelona, pp. 66–99.

Dalí, S. (1934) La visió de Gaudí. In: Descarnes, R., Prévost, C. and Pujols, F. (eds) *La Visió Artística i Religiosa de Gaudí*. Ayma, Barcelona.

Di Giovine, M. and Garcia-Fuentes, J.M. (2016) Sites of heritage, sites of pilgrimage. *International Journal of Tourism Anthropology* 5(1/2), 2–23.

Garcia-Fuentes, J.M. (2010) Tourism and Seductions of Difference. *Conference Proceedings, 1st International Conference, Tourism Contact Research Network*, Universidade de Lisboa, Lisbon, 2010.

Garcia-Fuentes, J.M. (2012) *The Construction of Modern Montserrat*. Universitat Politècnica de Catalunya, Barcelona.

Garcia-Fuentes, J.M. (2013) A nation of monasteries: the legacy of Víctor Balaguer in the Spanish conception of national monuments. *Future Anterior* 10(1), 40–51.

Garcia-Fuentes, J.M. (2015a) Reinventing and reshaping Gaudí: From nation and religion to tourism. Architecture, conflict and change in Barcelona's tourist imaginary. In: Graburn, N.N. and Gravari, M. (eds) *Tourism Imaginaries: Place, Practice, Media*. Ashgate, Surrey, UK, pp. 64–78.

Garcia-Fuentes, J.M. (2015b) A mountainous history of modern architecture? *Architectural Research Quarterly* 19(2), 102–105.

Garcia-Fuentes, J.-M. (2016) Guidebooks, postcards, and panoramas: the building of Montserrat through modern mass media. *Memory Studies* 9(1), 63–74. DOI: 10.1177/1750698015613973.

Garcia-Fuentes, J.M. (2018) The construction of modern Montserrat: architecture, politics and ideology. In: Jordan, K. and Lepine, A. (eds) *Modern Architecture and Religious Communities, 1850-1970: Building the Kingdom*. Routledge, London, pp. 49–69.

Garcia-Fuentes, J.M. (2019) Deconstructing Gaudí: entangled relationships between satire and architectural criticism. In: Rosso, M. (ed.) *Laughing at Architecture: Architectural Histories of Humour, Satire and Wit*. Bloomsbury Visual Arts, London and New York, pp. 91–108.

Garcia-Fuentes, J.M. and Garriga, S. (2012) Arquitectura y política en las exposiciones de Barcelona, 1888-1929. *Seminario Internacional sobre Eventos Mundiales y Cambio Urbano*, Grupo de Investigación HUM 700, Sevilla, pp. 72–79.

Guàrdia, M., Garcia Espuche, A. and Monclús, F.J. (1988) La dimensió urbana. In: Fabré i Carreras, X. (ed.) *Arquitectura i ciutat a l'exposició universal de Barcelona de 1888*. Edicions de la Universitat Politècnica de Catalunya, Barcelona, pp. 131–178.

Humboldt, W. von. (1803) Der Montserrat bey Barcelona. *Allgemeine Geographische Ephemeriden* 11, 265–313.

Humboldt, W. von. (1951) *Goethe y el Montserrat, Cuatro ensayos sobre Europa y América*, trad. Miguel de Unamuno. Espasa-Calpe, Madrid, pp. 117–153.

Junyent i Rafart, J. (1990) *Jaume Collell i Bancells: les campanyes patriótico-religioses (1878-1888)*. Patronat d'Estudis Ausonencs, Vic.

Lahuerta, J.J. (1999) *Antoni Gaudí (1852-1926). Arquitectura, ideología y política*. Electa, Madrid.

Lahuerta, J.J. (2016) *Antoni Gaudí: Ornament, Fire and Ashes*. Editorial Tenov, Barcelona.

Mabbett, I.W. (1983) The symbolism of Mount Meru. *History of Religions* 23(1), 64–83. DOI: 10.1086/462936.

Maragall, J. (1900) El templo que nace. *Diario de Barcelona* 20 (December), 10593–10594.

Maragall, J. (1905a) A la Sagrada Família. *Ilustració Catalana* 15 (January), 35–37.

Maragall, J. (1905b) ¡Una gr cia de caritat! *Diario de Barcelona* 7 (November), 12127–12128.

Maragall, J. (1907) Fuera del tiempo'. *Forma* 2(16), 134–147.

Marine-Roig, E. (2015) Religious tourism versus secular tourism: the Basilica of La Sagrada Família. *International Journal of Religious Tourism and Pilgrimage* 3(1), 25–37.

Mas, S. de. (1851) *La Iberia. Memoria sobre la conveniencia de la unión pacífica y legal de Portugal y España*. Imprenta y Estereotipia de M. Rivadeneyra, Madrid.

Montoliu, C. (1903) *Natura. Aplech d'estudis i descripcions de ses bellesas triats d'entre les obres de John Ruskin*. Joventut, Barcelona.

Olsen, D.H., Trono, A. and Fidgeon, P. (2018) Pilgrimage trails and routes: journeys from the past to the present. In: Olsen, D.H. and Trono, A. (eds) *Religious Pilgrimage Routes and Trails: Sustainable Development and Management*. CABI, Wallingford, UK, pp. 1–13.

Pi de Cabanyes, O. (2004) Víctor Balaguer i El *Risorgimento*. In: *Víctor Balaguer i el seu temps*. Publicacions de l'Abadia de Montserrat, Barcelona, pp. 11–39.

Pijoan, J. (1905) La Sagrada Família. La Catedral nova. *La Veu de Catalunya* 15(2334), 1.

Pijoan, J. (1906) ¡Acabem, acabem el temple! *La Veu de Catalunya* 16(2347), 3.

Roher, J. (1999) Una visió apropiada: El temple de la Sagrada Família de Gaudí i la política arquitectònica de la Lliga Regionalista. In: Lahuerta, J.J. (ed.) *Gaudí i el seu temps*. Barcanova, Barcelona, pp. 193–212.

Roma i Casanovas, F. (2002) *El paradís indicible: la representació de Montserrat durant l'època moderna*. Centre Excursionista de la Comarca del Bages, Manresa.

Roma i Casanova,F. (2004) *Del paradís a la nació. La muntanya a Catalunya. Segles XV-XX*. Cosset nia Edicions, Valls.

Timothy, D.J. (2013) Religious views of the environment: Santification of nature and implications for tourism. In: Holden, A. and Fennel, D. (eds) *The Routledge Handbook of Tourism and the Environment. Routledge*. Routledge, London, pp. 31–42.

Tomás y Valiente, T. (1985) *El Marco Político de la Desamortización en España*. Ariel, Barcelona.

Torii, T. (1983) *El mundo enigmático de Gaudí*. Instituto de España, Madrid.

Unamuno, M. and Gárate, J. (1951) *Glosas a Montserrat, Humboldt, W. von, Cuatro ensayos sobre Europa y América*, trad. Miguel de Unamuno. Espasa-Calpe, Madrid, pp. 150–153.

Zevi, B. (1950) *Storia dell'Architettura Moderna*. Einaudi, Turin.

9 Disease- and Health-related Issues at Mass Religious Gatherings

Daniel H. Olsen*
Department of Geography, Brigham Young University, Provo, Utah, United States

Introduction

Prior to the mid-20th century, travel was limited to a small segment of the human population, including seafarers, traders, military personnel, and religious pilgrims (Steffen, 2017a). However, with the global push towards neoliberalism and changes since 1960 in transportation technologies, there has been a steep increase in the mobility of people in many regions of the world (Olsen, 2019a). The United Nations World Tourism Organization (UNWTO) estimates that in 2018 there were 1.4 billion international tourists contributing 1.7 trillion USD to the global economy (UNWTO, 2019). With historical trends and projections suggesting ever-increasing numbers of international and domestic tourists, there are rising concerns by government and medical officials regarding how to maintain this growth while mitigating risks for tourists and residents of tourism destinations, particularly in the wake of escalating incidences of terrorism, natural disasters, and climate change impacts (e.g. Becken and Hay, 2007; Fuchs and Reichel, 2006; Korstanje and Clayton, 2012; Becken and Hughey, 2013).

Another area of concern is the health of both travellers and residents at tourism destinations and the creation of specialized services to meet the health needs of increasing numbers of travellers (Zuckerman, 2002). A growing tourism marketplace has led to the intensification of physical contact between people from different geographical areas, and therefore the potential spread of infectious and other types of diseases leading to global health security issues. This epidemiological concern is seen in the development of a specialized website by the World Health Organization (WHO) that specifically addresses issues related to international travel and health, including vaccinations and infectious diseases (https://www.who.int/ith/en/). The World Travel and Tourism Council (WTTC) has also expressed concerns regarding tourism and the potential spread of pandemics (WTTC, 2015), and presently co-chairs a 'Travel, Tourism, and Outbreaks' working group with the World Economic Forum and the WHO. These health-related concerns related to travel can also be seen in the plethora of research related to travel medicine, the establishment of specialised academic journals focused specifically on this issue, such as the *Journal of Travel Medicine* and *Travel Medicine and Infectious Disease*, and the development of professional and academic organizations such as the International Society of Travel Medicine (see www.istm.org).

At the same time, tourism has been viewed as a tool to increase people's physical, mental, emotional, and spiritual wellness as demonstrated by the growth of the health and

*dholsen@byu.edu

medical tourism niche market (e.g. Connell, 2006, Connell, 2011; Smith and Puczkó, 2008; Bushell and Sheldon, 2009; Hall, 2011, Voigt and Pforr, 2013). This has led to a 4.2 trillion USD global wellness economy that includes specialized spa treatments, fitness/mind and body, traditional and complementary medicine, preventative and personalized health programmes, and beauty and ageing niche markets (Global Wellness Institute, 2018). As such, tourism is ironically responsible for both increasing wellness and health among tourists with non-infectious diseases and the spreading of infectious diseases outside of their zone of influence.

While there is vast coverage of health-related issues at mass gatherings and to a smaller extent religious mass gathering in the travel medicine and risk management literature, this literature has not been accessed by scholars studying pilgrimage and religious tourism. To rectify this situation, the purpose of this chapter is to discuss the issue of diseases and other health-related issues as it concerns pilgrimage and religious tourism to mass religious events. First, this chapter reviews the academic literature on medicine, religion, and travel, and then discusses what constitutes a religious mass gathering. Then the chapter discusses the different types of infectious and non-infectious diseases and other health-risks that occur at religious mass gatherings with a particular emphasis on the *Hajj*. The chapter concludes with a discussion of how government, health, and religious officials are trying to mitigate the spread of diseases and other health-related issues at these religious mass gatherings and highlights the need for further research in this area.[1]

Medicine, Religion, and Travel

Ferngren (2014, pp. 4–6; c.f. Amundsen, 1996, pp. 1–29) notes several historical intersections between medicine and religion. The first is *medicine being subsumed by religion*. This relationship is found in homogeneous religious societies where disease and other health-related concerns and attempts at medical cures are shaped and understood through religious worldviews. The second is *where religion and medicine are partially separated*. In this case, societies 'understand disease [and other health issues] naturalistically and employ rational or empirical medicine for healing' (Ferngren, 2014, p. 5). However, questions related to the development of medical ethics are informed by religious worldviews and values. The third is where *religion and medicine have become completely separated and compartmentalized*. Ferngren notes that this is the situation in which most modern societies presently find themselves – where debates arise regarding the role of religion in medical decisions and ethics. As such, issues related to euthanasia, abortion, and assisted suicide become hot political and moral topics and are in many cases based on broader discussions regarding what role religion should play in not just these medical decisions but in society in general. The fourth is where *religion is subsumed by medicine*. In this instance, issues related to marriage, sex, mental health, and family relationships, which were once the purview of religion, are now being overtaken by a therapeutic perspective of health and healing. Ferngren argues that 'even spirituality has come to be viewed within a therapeutic framework, one in which counselors replace pastors, while the language becomes one of developing a healthy physical and spiritual lifestyle and restoring one's harmony with nature' (Ferngren, 2014, p. 5).

Another way in which medicine and religion are intertwined is through religiously-motivated travel. Indeed, pilgrimage, as one of the earliest motivations for travel (Timothy and Olsen, 2006; Butler and Suntikil, 2018), has long been intertwined with medicine and health-related concerns. For centuries, people have travelled to sacred places – both human-built and natural – seeking physical and spiritual healing. Indeed, some of the earliest therapeutic procedures involved taking a pilgrimage to a shrine of a god to petition for healing or relief from suffering (Gesler, 2003). This has been the case from Neolithic times, through the Graeco-Roman era, the Middle Ages, and down to the present day (Petsalis-Diomidis, 2006). During the Middle Ages, for example, Christians travelled in dangerous conditions to religious sites to venerate the relics of saints and petition them for healing and prosperity. Others travelled to monasteries where many monks offered some type of medicinal training or background (Scott, 2011; Ferngren and Lomperis, 2017). The popularity of pilgrimage in Europe during this time led to the creation of religiously-motivated medical services, ranging from the military order of Hospitallers (Knights of St John) to monastic hospitals (Ferngren and Lomperis, 2017). Religiously-motivated travel for healing purposes is also found historically in India and South America, where people sought out not only sacred places such as temples, shrines, and rivers, but holy men and women who were purported to have healing power (Ferrari, 2011; Palka, 2014). Today, pilgrims still travel to sacred sites to petition for physical and spiritual healing and continued health and prosperity for themselves and for others (Hufford, 1985; Gesler, 1996; Dubisch and Winkelman, 2005), and tend to take natural elements of sacred sites, such as holy water, soil, or flowers (Zeller, 1876; Shackley, 2006), home with them as holy relics (Stausberg, 2011; Farra-Haddad, Chapter 7, this volume).

Concerns regarding travel-related health issues pre-date modern transportation advances. As Steffen (2017a) notes, one of the first studies dealing with travel and epidemiology was conducted by James Lind (1716–1794) who did a controlled clinical trial of sailors as a part of his search for a cure for scurvy. Medical geographers have also been tracking the diffusion of disease and the rise of epidemics for decades (Meade and Emch, 2010). The modern rise of travel medicine occurred after World War II with increasing personal mobility and an emphasis on tourism development as an economic strategy (Steffen, 2017a). However, it was not until the 1970s that scholars and health officials began to gather and systematically assess epidemiological data. While accidents remain the most common cause of death among travellers (Steffen, 2017a), tourist exposure to infectious diseases at tourist destinations and their spread to other world regions is concerning to international governments and health officials due to the health and economic impacts of these diseases on tourist-generating and tourist-receiving regions. The 2003 SARS epidemic and its negative impacts on Asian tourism markets is a good example of this (e.g. McKercher and Chon, 2004; Cooper, 2005; Zeng *et al.*, 2005; Kuo *et al.*, 2008).

While exact numbers of tourists who develop a travel-related illness are unknown, some studies suggest that between 43–79% of travellers who frequently visit developing nations have acquired an illness (Angelo *et al.*, 2017a). Some of the common diseases that tourists both acquire and transmit during their travels include sexually transmitted diseases (De Schryver and Meheus, 1989; Memish and Osoba, 2006), respiratory tract infections (Cooper, 2005; Leder and Newman, 2005; Wilder-Smith, 2006), travellers diarrhoea (Harvey *et al.*, 2013; Steffen, 2017b), malaria (Spencer *et al.*, 2009; Angelo *et al.*, 2017b), hepatitis A and B (Löscher *et al.*, 1999; Keystone, 2005), skin infections (Hochedez *et al.*, 2009; Solomon *et al.*, 2011), and meningococcal disease (Memish, 2002). Other diseases are situational, such as outbreaks of SARS (Al-Tawfiq *et al.*, 2014), the zika virus (McCarthy, 2016; Petersen *et al.*, 2016), and dengue fever (Ratnam *et al.*, 2013) among others (see Zuckerman, 2002).

These illnesses, which occur because of or are facilitated through travel (Tatem *et al.*, 2012), have raised concerns over how to limit the transmission of these diseases as well as how to handle health issues related to infectious and non-infectious diseases, such as pulmonary disease, that may arise during the process of travelling (Dillard *et al.*, 1991; Mangili and Gendreau, 2005). Some strategies that have been employed to limit the transmission of pathogens or illnesses include travel warnings and restrictions (Weir, 2002), border screenings and requiring tourists to provide proof of having certain vaccinations before entering certain countries (St. John *et al.*, 2005), better environmental control systems on aircraft (Mangili and Gendreau, 2005), more education of tourists and tourism businesses (Van Herck *et al.*, 2004), the creation of global early warning systems (Yu and Madoff, 2004), and overall improved global health care systems (Wilder-Smith, 2006). In addition, there are several research areas that need to be addressed to improve travel medicine and restrict disease transmission, including tracking the changing geography of tourism as developing countries transition to become tourist-generating regions and understanding health problems that arise from travel between developing countries (Wilder-Smith, 2017).

Religious Mass Gathering Events

Mass gatherings, as the term suggests, occur when large numbers of people are gathered together in a particular geographic location to participate in an activity. However, there is a lack of consensus regarding a definition of these types of events due to disagreements regarding how large a crowd must be before a gathering or event can be given the moniker 'mass'. Baseline definitions of what is a mass gathering can range from as little as 1000 to over 25,000 or more (Arbon, 2004; Illiyas *et al.*, 2013). While defining what constitutes a mass gathering may seem a mundane and unimportant task, the development of a common definition of what constitutes a mass gathering is important in developing good management and medical practices and creating a consistent approach to data-gathering at these type of events (Arbon, 2004). Also, differentiating the different sizes of mass gathering is important in determining the resources required to run, as well as the time frame needed to plan for, the event. Molloy (2009, p. 268) suggests that that mass gatherings can be categorized into five types:

- *Small, medium, and large gatherings* range from 200-1,500 people (small), 1,500-10,000 people (medium), and 10,000-100,000 people (large) The resources for these gatherings are generally local in nature, although larger mass gatherings might also rely on county resources. The planning time to manage these gatherings has a range from 1 to 12 months.
- *Major mass gatherings* are gatherings of between 100,000 and 250,000 people. For these gatherings, regional and national resources are required.
- *Super mass gatherings* occur when between 250,000–500,000 people attend an event. National resources are required, and the planning process is generally greater than 12 months.
- *Extreme mass gatherings* are gatherings between 500,000–1,000,000 people. This type of gathering requires national and international resources to stage. The planning process for these events ranges between 12 and 24 months.
- *Mega mass gatherings* include gatherings of over 1,000,000 people. Like extreme mass gathering events, national and international resources are required, and the planning process is also between 12 and 24 months.

These categorizations are helpful when it comes to western tourism events planning, where religious passion plays and other pageants such as in Oberammergau, Germany, and North America (Chansky, 2006; Wetmore Jr, 2017; Olsen and Timothy, 2018); religious conferences and summer camps (Messenger, 1999); and religious pilgrimages, rituals, and processions to heavily-visited religious sites such as the Bascilica of Our Lady of Guadalupe in Mexico City, Notre Dame Cathedral in Paris, St Peter's Basilica in Vatican City, Rome, the Shrine of Padre Pio in San Giovanni Rotondo, Italy, the Western Wall in Jerusalem, the Meiji Shrine and Sensoki-Temple in Tokyo, and the Basilica de Sagrada Familia in Barcelona, can see hundreds of thousands of people participating in these events and millions of people visiting these sites on a yearly basis. In addition, several one-off events related to travel to see holy persons such as the Pope, and religious special events such as World Youth Day (Norman and Johnson, 2011; Gonzalez *et al.*, 2019) can also bring hundreds of thousands of people together in one location.

However, these numerical categorizations are rather meaningless when it comes to mass religious gatherings in non-western contexts. Indeed, religious mass gatherings in non-western countries constitute some of the largest mass gatherings in the world (Hsieh *et al.*, 2009). For example, the Kumbha Mela in India, which is celebrated every three years at four different locations, has approximately 120 million participants, and around 20 million people engage in the Arba'een Pilgrimage in Iraq annually. Therefore, a better indicator of categorizing mass religious gatherings might be relating the size of these gatherings to the daily average visitation and peak visitation numbers that are achieved on certain days during religious mass gatherings. For instance, birthdays or the anniversaries of the death of certain saints bring in a staggering number of devotees within Hindu religious traditions. The same can be said regarding full moon days in the Hindu and Muslim lunar calendars, the Jewish lunisolar calendar, and Christian calendars, where all pilgrimage sites see a large spike in pilgrim numbers (Singh, 2006; Olsen and Ron, 2013).

Types of Health Risks at Mass Religious Gatherings

In their research on health risks at the *Hajj*, Ahmed *et al.* (2006) categorize these risks at mass religious events into two main categories: the contraction of communicable diseases and health risks related to pre-existing non-communicable diseases. This section discusses these two categories and adds a third category under the heading of 'other' types of health-related risks at mass religious events.

Communicable diseases

While international travel and commerce are the main factors contributing to the development and spread of emerging and reemerging pathogens in the modern era (Ostroff and Kozarsky, 1998), mass events, such as music and sporting events, have also been associated with the spread of disease (e.g. Feldman *et al.*, 2004; Hadjichristodoulou *et al.*, 2005; Eberhardt *et al.*, 2016). This is the case with religious mass gatherings that involve millions of local, regional, and international pilgrims, religious tourists, and tourists converging on a particular geographic point to participate in religious festivals and other activities.

The sheer number of participants at these religious mass gatherings can present several management and logistical challenges, including dealing with the potential health issues that come with this temporal and spatial concentration of people. This is in part because of the geographical proximity of people that causes these mass religious gatherings to become 'hot spots' for the potential contraction and spread of communicable or infectious diseases (Steffen *et al.*, 2012) through direct or indirect person to person contact. For example, diseases can be transmitted by an infected person shaking the hands of another person or by an infected person touching an environmental surface in a high traffic area, droplets, aerosolized pathogens, and contaminated food and water. In addition, communicable diseases can also be transmitted through abiotic factors, such as wind (e.g. inhaling spores), or animal vectors, such as mosquitoes and fleas. Within this category, Ahmed *et al.* (2006) suggest several subcategories of communicable diseases that have been found present at the *Hajj*, including respiratory tract infections, skin infections, blood-borne diseases, and the introduction of emerging or new infectious diseases.

When participants at religious mass events become vectors for the transmission of these diseases, they unwittingly transmit them back to their pilgrim-generating country and even beyond. This spread of communicable diseases by pilgrims to their home regions generally occurs due to the lack of monitoring of pilgrims who travel abroad. Because pilgrims and religious tourists can travel anywhere in the world in a matter of days, and some incubation periods can last weeks, the manifestation of symptoms can occur after a tourist has returned home. For example, Rondy *et al.* (2011) note that in October 2008 there was an outbreak of norovirus in a psychiatric institution in The Netherlands that occurred after some of the patients returned from a pilgrimage to Lourdes. Another example comes from the *Hajj*, where in 2000 there were 90 instances of meningococcal infection transmitted from pilgrims who had completed the *Hajj*. While the people infected came from nine countries, most of the instances were found in England and France. A total of 14 people died, including family members of the pilgrims and others who had been in contact with the pilgrim but had no relationship to them (Aguilera *et al.*, 2000).

In other cases, infectious diseases can be carried by pilgrims from pilgrim-generating regions to pilgrimage destinations. This is why the Saudi Arabian government will not issue *Hajj* visas to pilgrims unless they have proof of current vaccinations for common infectious diseases, such as yellow fever, polio, and seasonal influenza (Ahmed *et al.*, 2006). In addition, residents in towns located along pilgrimage routes can also be exposed to diseases carried by pilgrims. Shinde, 2018 illustrates this in his research on the Dnyaneshwar palkhi or pilgrimage in India that involves half a million people walking 210 km over 14 days as they travel from Alandi to Pandharpur. The palkhi stops for the evening at one of the 14 towns along the pilgrimage path. Because of the lack of infrastructure (e.g. toilets, clean drinking water, and waste disposal), pilgrims leave behind vast amounts of garbage and human waste that, if not disposed of quickly and properly, can lead to illness among the town residents.

Non-communicable diseases

The second category of health risks are non-communicable diseases that are chronic conditions brought by attendees to religious mass events. Since non-communicable disease is manifest in a person as a result of a combination genetic, environmental, hereditary, and physiological factors, such diseases cannot be spread to other attendees. At all mass events, including religious mass events, the incidence of acute cardiovascular health-related issues more than doubles, particularly in instances where intense emotional stress is present, such as sporting events or events with severe overcrowding (Steffen *et al.*, 2012). Ahmed *et al.* (2006) highlight cardiovascular diseases as a concern for *Hajj* attendees, but other types of non-communicable diseases that can manifest themselves or be exacerbated within participants at religious mass gatherings include autoimmune diseases, lung diseases, cancer, cataracts, and osteoporosis, which in some cases can lead to morbidity, disability, and even death.

Other Health-related risks

While discussed under the category of non-communicable diseases by Ahmed *et al.* (2006), for the purposes of this chapter this final category of health risks is labelled as 'other health risks' – those that are present at large religious gatherings. These health risks include trauma risks (e.g. human stampedes, accidents), fire-related injuries (e.g. cooking-related injuries), environmental heat injuries (e.g. heat exhaustion and stroke), and occupational hazards to those who work at these events. This category of 'other' that involves environmental and accidental health risks is seen by experts to be the major cause of mortality at mass gatherings (Steffen *et al.*, 2012) – that is, more people die from accidents, overcrowding, and heatstroke at religious mass events than communicable and non-communicable diseases. This is because it is a dense concentration of people in a particular location that elevates the potential for environmental health concerns such as overcrowding, human stampedes, and heat-related illnesses (Steffen *et al.*, 2012; see Illiyas *et al.*, 2013; Gayathri *et al.*, 2017). However, it is the diffusion of disease to other regions after mass religious gatherings that tends to worry health and government officials the most. The indirect impacts of prolonged mass religious gatherings, such as pollution from traffic congestion, can cause respiratory issues, and waste that is generated from the materials used in ritual offerings that in turn create breeding grounds for bacteria (see Chapter 3, this volume), are also causes for concern.

Disease and Health Risks at the *Hajj*

Every year, between 2–3 million Muslims travel to Saudi Arabia to participate in the *Hajj*. This pilgrimage is a requirement for all able-bodied Muslims who have the funds to participate. While many pilgrims come from the Middle East region, a growing number come from Africa, Asia, Europe, the Americans, and Australia. This makes the *Hajj* one of the largest, as well as one of the oldest, annual mass transnational human migrations in the world (Almutairi *et al.*, 2018; Timothy and Iverson, 2006). The number of people who participate in the *Hajj* requires an incredible amount of management and logistical planning, including maintaining accommodation and supporting infrastructure (Henderson, 2011); crowd management and issues related to extreme congestion and pilgrimage flows (Charnes *et al.*, 1989; Johansson *et al.*, 2012; Yamin and Albugami, 2014); disaster management (Manoochehry and Rasouli, 2017); transportation management (Tayan *et al.*, 2014); and waste management (Johansson *et al.*, 2012;

Anjum *et al.*, 2016; see Chapter 10, this volume) among others.

As Gautret and Steffen (2016) note, most academic papers related to mass religious gatherings, diseases, and health-related issues have focused on the *Hajj* and *Umrah* – the *Umrah* being a smaller pilgrimage which can be completed at any time during the year. This is due in part because of the long-standing interest by European powers in both the region and the host of epidemics that spread from *Hajj* pilgrims – as vectors of disease – to Europe and other world regions. For example, in 1865, a massive cholera epidemic occurred in Mecca that corresponded to a cholera outbreak in several European countries that same year. This led European colonial powers to medicalize the *Hajj* through policies that, while subsumed under larger political concerns about 'Islamism', 'fashioned [and expanded] medical practices and debates' (Mishra, 2011, p. 149) at the *Hajj*, where stringent European sanitary and medical practices and regulations regarding who could attend were enforced. As such, the 'medicalization' of the *Hajj*, which also included the international oversight of *Hajj* management and the regimentation of medical practices, the forced vaccination of *Hajj* pilgrims, the move from quarantine to 'disinfection and surveillance' (Huber, 2016), the on-the-spot screening of pilgrims' bodies, and the health data gathered for analysis over the decades (Huber, 2016), has allowed for more detailed health studies and the development of a more robust research agenda regarding disease and other health issues at the *Hajj*.

Several hundred academic papers have been published focusing on the identification and medical management of communicable diseases that are present and contracted by pilgrims at the *Hajj* (Table 9.1). Of particular concern among scientists are diseases that affect pilgrims' respiratory systems. As Al-Ghamdi *et al.* (2003) note, the number one reason *Hajj* pilgrims are admitted to hospitals are respiratory system issues, followed by cardiovascular issues and gastrointestinal tract issues. Of these health problems, pneumonia, asthma, and chronic obstructive pulmonary diseases were the most common, most of which were due to pre-existing conditions (see Baharoon *et al.*, 2009). In addition to focusing on how *Hajj* officials dealt with communicable disease outbreaks, other papers concentrated on the impacts of diseases carried by *Hajj* pilgrims back

Table 9.1. A table showing the different types of communicable diseases present at the *Hajj* as discussed in the academic/medical literature (Ahmed *et al.*, 2006; Choudhry *et al.*, 2006; Gatrad *et al.*, 2006; Shafi *et al.*, 2008; Baharoon *et al.*, 2009; Rafiq *et al.*, 2009; Memish, 2010; Moattari *et al.*, 2012; Al-Tawfiq *et al.*, 2013; Haworth *et al.*, 2013; Memish *et al.*, 2014; Yezli *et al.*, 2016).

Disease
Avian influenza
Blood-borne diseases
Hepatitis B and C
HIV
Diarrhoeal disease
Cholera
Food poisoning
Hepatitis A
Gastrotinestinal infections
Meningococcal disease
Polio virus
Respiratory tract infections
Haemophilus influenza
Middle East respiratory syndrome coronavirus (MERS-CoV)
Influenza
Klebsiella pneumoniae
Mycobacterium tuberculosis
Parainfluenza
Pertussis
Respiratory syncytial virus (RSV)
Streptococcus pneumoniae
Sepsis and septic shock
Severe acute respiratory syndrome (SARS)
Skin infections
Vector-borne diseases
Arboviruses
Malaria
Viral haemorrhagic fever (VHF) syndrome
Yellow fever
Zoonotic diseases

to their countries of origin. For example, the outbreak of W135 meningococcal disease following the 2000 and 2001 *Hajj* in Europe and Mauritius was a topic of intensive study (Taha *et al.*, 2000; Aguilera *et al.*, 2000; Hahné *et al.*, 2002; Issack and Ragavoodoo, 2002), as was research on the 2009 outbreak of influenza, influenza A H3N2, or pandemic H1N1, transmitted by *Hajj* pilgrims in the Fars province in Iran and in France (Gautret *et al.*, 2009, 2013; Moattari *et al.*, 2012).

With regard to non-communicable diseases, Al-Ghamdi *et al.* (2003) note that the most prevalent pre-existing medical conditions that were treated were bronchial asthma, chronic obstructive pulmonary disease, hypertension, and diabetes mellitus; while Gautret and Steffen (2016) also show that hypertension and hypercholesterolemia are also common problems among *Hajj* attendees. Yassen and Sameer (2006) also note that cerebrovascular diseases are a common cause of death among *Hajj* attendees. In terms of other health-related issues among pilgrims at the *Hajj*, environmental heat-related injuries (heatstroke and heat exhaustion, fire-related injuries, sunburn, dehydration, slaughter-related and abattoir worker injuries) and injuries related to motor vehicle accidents, human stampedes, and accidental injuries (e.g. broken bones and sprained muscles and joints) are common (Ahmed *et al.*, 2006). Human stampedes in particular have been a concern at the *Hajj*, where there are several instances of hundreds of people being trampled at choke and exit points (e.g. Ahmed *et al.*, 2006; Memish *et al.*, 2012; Salamati and Rahimi-Movaghar, 2016; Alaska *et al.*, 2017). Not only can these events lead to death, but general overcrowding can also delay sanitation and waste servicing and cause further breeding of bacteria causing respiratory and other communicable diseases (Yamin, 2015).

Disease and Health Risks at Non-*Hajj*/*Umrah* Mass Religious Gatherings

In most cases, the same types of disease and health risks found at the *Hajj* and the *Umrah* pilgrimage events are also found at other religious mass gatherings. For example, in the case of other Muslim pilgrimages, Gautret and Steffen (2016) document instances of febrile illness (i.e. sudden illnesses accompanied with a fever) at the 2010 annual celebration of the Day of Ashura in Karbala, Iraq; acute respiratory illness and diarrhoea at the 2010 annual celebration of the Urs of Baba Farid in Pakpattan, Pakistan; and gastrointestinal disorders at the 2009 and 2010 Moulay Abdellah Amghar Moussem in Morocco. As Gautret and Steffen note, no systematic assessment of these outbreaks could be made because of the lack of data. Sokhna *et al.* (2017), in their study of the Muslim pilgrimage gathering at the Grand Magal of Touba in Senegal, found that the most frequent requests for medical aid were related to trauma injuries, fatigue, and heatstroke. However, there was also a high rate of infectious disease, including febrile systemic illnesses, malaria, diarrhoeal diseases, and respiratory tract infections.

In their review of non-*Hajj* and -*Umrah* mass religious gatherings, Gautret and Steffen (2016) note that certain pilgrimage and other religious gatherings – at least the ones where health statistics were available – show that the most common communicable disease outbreaks at these gatherings included measles, influenza, mumps, hepatitis A and gastrointestinal infections. Interestingly, these diseases are vaccine-preventable and could be minimized through immunization. At the 2010 Holy Shroud exhibition in Torino, Italy, Bortolin *et al.* (2013) note that cardiac and trauma emergencies were the most common health issue during the event. Also, at the Kumbha Mela, David and Roy (2016) found that human bathing in the nearby rivers led to an increased incidence of waterborne infections and diseases. However, outside occasional cholera outbreaks, there is little data regarding infectious diseases and health risks at these Kumbha Mela events, possibly due to the lack of fixed pilgrimage settlements and health facilities seen at the *Hajj*. Instead, pilgrims create spontaneous and ad-hoc settlements that vary from year to year, causing health officials to service these settlements in an improvised manner. As such, there is little reliable data regarding the heath issues that arise among pilgrims.

The Management and Prevention of Health Risks at Mass Religious Gatherings

With the increasing popularity of travel to religious heritage sites and cultural events, scholars have become interested in management strategies regarding how to handle increasing numbers of visitors at these locations (e.g. Shackley, 2001a,b, 2003; Woodward, 2004; Olsen, 2006, Olsen, 2019b; Henderson, 2011; Olsen and Ron, 2013; Bond *et al.*, 2015). The same can be said in the context of religious mass gatherings. In particular, as noted above, because of the decades-worth of health planning at the *Hajj*, much of the academic literature on health-related issues at religious mass gatherings has focused on how authorities responsible for the *Hajj* have tried to mitigate the spread of communicable diseases, care for people with non-communicable diseases, and lessen incidences of stampeding and heat, fire, and accident-related injuries. Indeed, because the *Hajj* is held yearly in the same location, authorities and officials have developed over time a dedicated medical infrastructure with the help of international organizations and developed countries.

However, in many cases, religious gatherings are one-off special events or are held infrequently, and therefore organizers of these gatherings lack a strategic plan regarding health services, and so the delivery of those services suffers from inadequate planning and inefficient surveillance systems used to monitor public health risks. There is often poor crowd management and coordination and a lack of basic sanitation services, all of which can intensify health risks (Abubakar *et al.*, 2012). Also, infrastructure, climate and weather, sex, age, crowd density, and mood seem to determine both the types and the intensity of health risks at these gatherings (Steffen *et al.*, 2012). If these elements are not considered in the planning process, then health-related issues at these events are only exacerbated. However, based on the available research, particularly as it relates to the *Hajj*, this section focuses on best practices regarding how governments, religious organizations, and health authorities are attempting to monitor and limit instances of health-related issues at religious mass events, including mandatory vaccinations, the expansion and coordination of modern health facilities and practices, and the use of technology to monitor and control disease outbreaks and stampedes.

Mandatory vaccinations

Government-mandated vaccinations have been in place for several decades because of several outbreaks of disease at the *Hajj* and among return pilgrims. In 1987, a large outbreak of meningococcal disease serogroup A occurred among Hajj attendees. This prompted the Saudi Arabian health authorities to implement a series of preventative strategies including compulsory vaccination with bivalent A and C vaccine for all arriving pilgrims, vaccination campaigns in countries sending a high number of *Hajj* participants, and compulsory oral ciprofloxacin for pilgrims coming from the sub-Sahara Africa region (Ahmed *et al.*, 2006). Outbreaks of other strains of meningococcal disease were frequent among *Hajj* attendees until 2002, when the Saudi government required attendees to provide proof of quadrivalent meningococcal vaccination before receiving a *Hajj* visa (Steffen, 2010). In 2013, mandatory vaccinations for pilgrims included vaccinations for yellow fever, meningitis, and polio (Memish *et al.*, 2014).

In addition to these required vaccinations, *Hajj* and other government officials coordinate with *Hajj*-sending countries to require attendees to have received other vaccinations based on recommendations by the WHO and the US Centers for Disease Control and Prevention (CDC), organizations that engage in ongoing emerging disease monitoring and surveillance through updating global health data and tracking global outbreaks (Al-Harthi and Al-Harbi, 2001; Memish *et al.*, 2014). Every year, the Saudi government releases the health requirements that need to be met for *Hajj* pilgrims to receive a travel visa (Memish, 2010; Memish *et al.*, 2014; Shafi *et al.*, 2016). As a part of this move towards mandatory vaccinations, in 2012 the Saudi Ministry of Health established the Global Center for Mass Gathering Medicine, which networks with the WHO Collaborating Centre on Mass Gatherings

Medicine, the WHO Department of Global Capacity Alert and Response, and several academic and public health institutions in the UK in the creation of these vaccination standards.

Expanding and coordinating health facilities

At many religious mass gatherings, health issues arise because of overcrowding, climatic conditions, and the lack of sanitary facilities and medical resources (Sokhna *et al.*, 2017). However, *Hajj* officials have developed an effective and efficient system to handle health-related issues through '...rapid diagnosis, timely treatment, prevention by vaccine, community measures, infection prevention and control practices' (Al-Harthi and Al-Harbi, 2001, p. 28). At the *Hajj*, medical facilities and personnel are well equipped to handle many of the health issues that arise during the pilgrimage event. While the majority of mass gathering injuries and health complaints can be effectively treated at the scene (Olapade-Olaopa *et al.*, 2006), Memish (2010) notes that the Saudi government has 136 health care centres located near the main gathering places of the *Hajj* that are staffed by over 17,000 medical personnel. Also, along the pilgrimage trail the Saudi Ministry of Health, with the help of the National Guard, the Internal Security Forces, and the Ministry of Defence and Aviation, offer 24-hour medical care to the pilgrims. At the King Abdulaziz International Airport in Jeddah where most *Hajj* pilgrims arrive, there are two clinical examination rooms that provide health care, including the administration of recommended prophylactic medicines. This health care is provided to pilgrims free of charge (Memish, 2010). In fact, the Saudi government provides free health care for the pilgrims during the entire *Hajj* pilgrimage (Memish, 2010).

Saudi government and health officials also engage in extensive pre-*Hajj* public preparedness projects and interventions. For example, in addition to pre-travel health regulations and information regarding vaccinations, the Saudi Ministry of *Hajj* and the Ministry of Health do thorough inspections and testing of water sources at holy sites as well as sewage treatment plants throughout the year. Additionally, those that work in the food sector are tested for pathogens, and hotels and hospitals are also tested to maintain standards of food hygiene. The government has wide vector control and insecticide spraying programmes, including spraying the tents provided for pilgrims with long-lasting insecticide and aerial spraying over Arafat and Mina (Memish, 2010).

Technology use

While modern health technologies are utilized by *Hajj* officials, issues related to the number of injuries related to human stampedes and overcrowding are still a concern. Even with an increase in the number of security cameras used by *Hajj* officials to monitor pilgrim movements (Alabdulkarim *et al.*, 2016), several researchers have offered suggestions with regards to different types of technologies used to mitigate these health-related concerns. For example, Yamin (2015) suggests that the use of radio frequency identification (RFID) technology embedded in pilgrim wristbands or in pilgrims' phones (David and Roy, 2016) rather than in RFID-embedded cards issued to pilgrims by the Saudi government would act as more efficient personal GPS tracking devices, allowing authorities to better follow pilgrim flows and to provide more efficient support for medical emergencies in real time (see and Ali *et al.*, 2015 and Alqahtani *et al.*, 2016). Batty (2006) has advocated the use of simulations of pilgrim crowding at key choke points such as the Jamarat Bridge, a pedestrian bridge that allows pilgrim access to the ritual stoning of the devil during a key point in the *Hajj* pilgrimage ritual. Alabdulkarim *et al.* (2016) suggest that the use of unmanned aerial vehicles and mobile crowd sensing and computing technology will also facilitate crowd control through better crowd detection and flow monitoring, thereby creating faster response times by health-responders.

Conclusion

The purpose of this chapter was to provide a brief overview of the ties between medicine, health, pilgrimage, and religious tourism. More particularly, focus was on mass religious gatherings where, like other large gatherings of people, there are several health risks relating to the spread of communicable diseases, dealing with people who participate in events with pre-existing non-communicable diseases, and other health-related issues related to overcrowding, environmental factors, and injuries. While many of these health issues stem from poor management and planning, including the lack of hygienic facilities, the close proximity of people in sometimes small, confined spaces also exacerbates the rates of contraction and the spread of communicable diseases as well as other injury risks. As such, there have been several studies looking at how to minimize these health risks at large religious gatherings, in particular during the *Hajj*.

Most of the discussion in this chapter has focused on the *Hajj*, in part because of its long history as a case study for mass gathering medicine and concerns over global health security from pilgrims returning from the *Hajj*. However, some of the management techniques are also used at larger religious mass gatherings. For example, mandatory vaccinations are historically a part of the Kumbha Mela festival and were instituted by the Indian government after Asiatic cholera pandemics originating from Kumbha Mela pilgrims to their countries of origin were observed in 1892, 1948, and in the 1960s. However, the government has recently relaxed the regulations for mandatory vaccinations and is instead focusing on addressing a wider range of pilgrim health needs, including enhanced medical services and cleaner food and water access (David and Roy, 2016). However, there is still a lack of an international and interdisciplinary approach to meeting the health challenges at the Kumbha Mela festival, and more broadly a need for a greater 'multidisciplinary international association' to study and develop improved policies and practices at religious mass gatherings outside of the *Hajj* (David and Roy, 2016).

In addition, the *Hajj* is a large-scale international event that appears to have little to no application to smaller religious mass gatherings (e.g. below 100,000), where coordination of governments, health officials, and academic institutions may not be necessary, and even then only on a local or regional level. The use of new technologies, such as RFID chips and unmanned aviation equipment may help with crowding and mitigating human stampedes. However, this technology is expensive and not readily available for organizers of smaller religious mass gatherings. There is a need at all scales to be more aware of the emerging infectious diseases at a global level and to engage in preventative measures for incoming international pilgrims and religious tourists in alignment with appropriate travel visa requirements. In the case of religious gatherings where over 100,000 or more people are expected to attend, many of whom attend for healing or for relief from physical suffering, there is a pressing need for continued communication and coordination with health officials, necessitating the need for more interagency cooperation at all levels of government.

Note

[1] Endnote 1: This chapter was written and submitted for publication prior to the 2020 COVID-19 pandemic. This pandemic highlights some of the concerns governments and health officials have long held regarding travel to more generally. In the wake of this pandemic, religious pilgrims sites have been closed and pilgrimage event either been canceled or amended to cater to the requests by government and health officials for social distancing. While it is probably too soon to see how this pandemic affects religious pilgrimage and tourism to religious sites and also mass religious gatherings, the author imagines that pilgrimage and religious tourism may help to kick start tourism, at least domestically in the next couple of years as people who have sought after God during the pandemic will want to travel to sacred sites to give thanks or to participate in religious rituals for health purposes.

References

Abubakar, I., Gautret, P., Brunette, G.W., Blumberg, L., Johnson, D. *et al.* (2012) Global perspectives for prevention of infectious diseases associated with mass gatherings. *The Lancet Infectious Diseases* 12(1), 66–74. DOI: 10.1016/S1473-3099(11)70246-8.

Aguilera, J.-F., Perrocheau, A. and Meffre, C. (2000) Hahné, S. and W135 Working Group, 2002. Outbreak of serogroup W135 meningococcal disease after the Hajj pilgrimage, Europe. *Emerging Infectious Diseases* 8(8), 761–767.

Ahmed, Q.A., Arabi, Y.M. and Memish, Z.A. (2006) Health risks at the Hajj. *The Lancet* 367(9515), 1008–1015. DOI: 10.1016/S0140-6736(06)68429-8.

Al-Ghamdi, S.M., Akbar, H.O., Qari, Y.A., Fathaldin, O.A. and Al-Rashed, R.S. (2003) Pattern of admission to hospitals during Muslim pilgrimage (Hajj). *Saudi Medical Journal* 24(10), 1073–1076.

Al-Harthi, A.S. and Al-Harbi, M. (2001) Accidental injuries during Muslim pilgrimage. *Saudi Medical Journal* 22(6), 523–525.

Al-Tawfiq, J.A., Zumla, A. and Memish, Z.A. (2013) Respiratory tract infections during the annual Hajj: potential risks and mitigation strategies. *Current Opinion in Pulmonary Medicine* 19(3), 192–197.

Al-Tawfiq, J.A., Zumla, A. and Memish, Z.A. (2014) Travel implications of emerging coronaviruses: SARS and MERS-CoV. *Travel Medicine and Infectious Disease* 12(5), 422–428. DOI: 10.1016/j.tmaid.2014.06.007.

Alabdulkarim, L., Alrajhi, W. and Aloboud, E. (2016) Urban analytics in crowd management in the context of Hajj. In: Meiselwitz, G. (ed.) *Social Computing and Social Media*. Springer, Cham, Switzerland, pp. 249–257.

Alaska, Y.A., Aldawas, A.D., Aljerian, N.A., Memish, Z.A. and Suner, S. (2017) The impact of crowd control measures on the occurrence of stampedes during mass gatherings: the Hajj experience. *Travel Medicine and Infectious Disease* 15, 67–70. DOI: 10.1016/j.tmaid.2016.09.002.

Ali, M.F., Bashar, A. and Shah, A. (2015) SmartCrowd: Novel Approach to Big Crowd Management using Mobile Cloud Computing. In *2015 International Conference on Cloud Computing (ICCC)* (pp. 229-232), IEEE, Piscataway, New Jersey. Available at: https://ieeexplore.ieee.org/xpl/conhome/7145509/proceeding (accessed 2 July 2020).

Almutairi, M.M., Alsalem, W.S., Hassanain, M. and Hotez, P.J. (2018) Hajj, Umrah, and the neglected tropical diseases. *PLOS Neglected Tropical Diseases* 12(8), e0006539. DOI: 10.1371/journal.pntd.0006539.

Alqahtani, A.S., BinDhim, N.F., Tashani, M., Willaby, H.W., Wiley, K.E. *et al.* (2016) Pilot use of a novel smartphone application to track traveller health behaviour and collect infectious disease data during a mass gathering: Hajj pilgrimage 2014. *Journal of Epidemiology and Global Health* 6(3), 147–155. DOI: 10.1016/j.jegh.2015.07.005.

Amundsen, D.W. (1996) *Medicine, Society, and Faith in the Ancient and Medieval Worlds*. John Hopkins University Press, Baltimore, Maryland.

Angelo, K.M., Kozarsky, P.E., Ryan, E.T., Chen, L.H. and Sotir, M.J. (2017a) What proportion of international travellers acquire a travel-related illness? A review of the literature. *Journal of Travel Medicine* 24(5), 1–8. DOI: 10.1093/jtm/tax046.

Angelo, K.M., Libman, M., Caumes, E., Hamer, D.H., Kain, K.C. *et al.* (2017b) Malaria after international travel: a GeoSentinel analysis, 2003–2016. *Malaria Journal* 16(1), 293. DOI: 10.1186/s12936-017-1936-3.

Anjum, M., Miandad, R., Waqas, M., Ahmad, I., Alafif, Z.O.A. *et al.* (2016) Solid waste management in Saudi Arabia. *Journal of Applied Agriculture and Biotechnology* 1, 13–26.

Arbon, P. (2004) The development of conceptual models for mass-gathering health. *Prehospital and Disaster Medicine* 19(3), 208–212. DOI: 10.1017/S1049023X00001795.

Baharoon, S., Al-Jahdali, H., Al Hashmi, J., Memish, Z.A. and Ahmed, Q.A. (2009) Severe sepsis and septic shock at the Hajj: etiologies and outcomes. *Travel Medicine and Infectious Disease* 7(4), 247–252. DOI: 10.1016/j.tmaid.2008.09.002.

Batty, M. (2006) Agent-based technologies and GIS: simulating crowding, panic, and disaster management. In: Rana, S. and Sharma, J. (eds) *Frontiers of Geographic Information Technology*. Springer, Berlin, pp. 81–101.

Becken, S. and Hay, J.E. (2007) *Tourism and Climate Change: Risks and Opportunities*. Channel View Publications, Bristol, UK.

Becken, S. and Hughey, K.F.D. (2013) Linking tourism into emergency management structures to enhance disaster risk reduction. *Tourism Management* 36, 77–85. DOI: 10.1016/j.tourman.2012.11.006.

Bond, N., Packer, J. and Ballantyne, R. (2015) Exploring visitor experiences, activities and benefits at three religious tourism sites. *International Journal of Tourism Research* 17(5), 471–481. DOI: 10.1002/jtr.2014.

Bortolin, M., Ulla, M., Bono, A., Ferreri, E., Tomatis, M. *et al.* (2013) Holy Shroud exhibition 2010: health services during a 40-day mass-gathering event. *Prehospital and Disaster Medicine* 28(3), 239–244. DOI: 10.1017/S1049023X13000216.

Bushell, R. and Sheldon, P.J. (eds) (2009) *Wellness and Tourism: Mind, Body, Spirit, Place*. Cognizant Communication, New York.

Butler, R. and Suntikil, W. (2018) *Tourism and Religion: Issues, Trends and Implications*. Channel View Publications, Bristol, UK.

Chansky, D. (2006) North American Passion Plays: "The Greatest Story Ever Told" in the New Millenium. *TDR/The Drama Review* 50(4), 120–145. DOI: 10.1162/dram.2006.50.4.120.

Charnes, A., Duffuaa, S. and Yafi, A. (1989) A non-linear congestion network model for planning internal movement in the Hajj. *European Journal of Operational Research* 40(3), 292–298. DOI: 10.1016/0377-2217(89)90421-9.

Choudhry, A.J., Al Mudaimegh, K.S., Turkistani, A.M. and Al Hamdan, N.A. (2006) Hajj-associated acute respiratory infection among Hajjis from Riyadh. *EMHJ - Eastern Mediterranean Health Journal,* 12(3-4), 300–309.

Connell, J. (2006) Medical tourism: Sea, sun, sand and ... surgery. *Tourism Management* 27(6), 1093–1100. DOI: 10.1016/j.tourman.2005.11.005.

Connell, J. (2011) *Medical Tourism*. CAB International, Wallingford, UK.

Cooper, M. (2005) Japanese tourism and the SARS epidemic of 2003. *Journal of Travel & Tourism Marketing* 19(2-3), 117–131. DOI: 10.1300/J073v19n02_10.

David, S. and Roy, N. (2016) Public health perspectives from the biggest human mass gathering on earth: Kumbh Mela, India. *International Journal of Infectious Diseases* 47, 42–45. DOI: 10.1016/j.ijid.2016.01.010.

De Schryver, A. and Meheus, A. (1989) International travel and sexually transmitted diseases. *World Health Statistics Quarterly* 42(2), 90–99.

Dillard, T.A., Beninati, W.A. and Berg, B.W. (1991) Air travel in patients with chronic obstructive pulmonary disease. *Archives of Internal Medicine* 151(9), 1793–1795. DOI: 10.1001/archinte.1991.00400090085015.

Dubisch, J. and Winkelman, M. (eds) (2005) *Pilgrimage and Healing*. University of Arizona Press, Tuscon.

Eberhardt, K.A., Vinnemeier, C.D., Dehnerdt, J., Rolling, T., Steffen, R. *et al.* (2016) Travelers to the FIFA World Cup 2014 in Brazil: health risks related to mass gatherings/sports events and implications for the summer Olympic Games in Rio de Janeiro in 2016. *Travel Medicine and Infectious Disease* 14(3), 212–220. DOI: 10.1016/j.tmaid.2016.05.014.

Feldman, M.J., Lukins, J.L., Verbeek, P.R., MacDonald, R.D., Burgess, R.J. *et al.* (2004) Half-a-million strong: the emergency medical services response to a single-day, mass-gathering event. *Prehospital and Disaster Medicine* 19(04), 287–296. DOI: 10.1017/S1049023X00001916.

Ferngren, G.B. (2014) *Medicine & Religion: A Historical Introduction*. John Hopkins University Press, Baltimore.

Ferngren, G.B. and Lomperis, E.N. (2017) *Essential Readings in Medicine & Religion*. John Hopkins University Press, Baltimore.

Ferrari, F. (ed.) (2011) *Health and Religious Rituals in South Asia: Disease, Possession and Healing*. Routledge, London and New York.

Fuchs, G. and Reichel, A. (2006) Tourist destination risk perception: the case of Israel. *Journal of Hospitality & Leisure Marketing* 14(2), 83–108. DOI: 10.1300/J150v14n02_06.

Gatrad, A.R., Shafi, S., Memish, Z.A. and Sheikh, A. (2006) Hajj and the risk of influenza. *BMJ* 333(7580), 1182–1183. DOI: 10.1136/bmj.39052.628958.BE.

Gautret, P., Soula, G., Delmont, J., Parola, P. and Brouqui, P. (2009) Common health hazards in French pilgrims during the Hajj of 2007: a prospective cohort study. *Journal of Travel Medicine* 16(6), 377–381. DOI: 10.1111/j.1708-8305.2009.00358.x.

Gautret, P., Parola, P. and Brouqui, P. (2013) Relative risk for influenza like illness in French hajj pilgrims compared to non-Hajj attending controls during the 2009 influenza pandemic. *Travel Medicine and Infectious Disease* 11(2), 95–97. DOI: 10.1016/j.tmaid.2013.03.003.

Gautret, P. and Steffen, R. (2016) Communicable diseases as health risks at mass gatherings other than Hajj: what is the evidence? *International Journal of Infectious Diseases* 47, 46–52. DOI: 10.1016/j.ijid.2016.03.007.

Gayathri, H., Aparna, P.M. and Verma, A. (2017) A review of studies on understanding crowd dynamics in the context of crowd safety in mass religious gatherings. *International Journal of Disaster Risk Reduction* 25, 82–91. DOI: 10.1016/j.ijdrr.2017.07.017.

Gesler, W. (1996) Lourdes: healing in a place of pilgrimage. *Health & Place* 2(2), 95–105. DOI: 10.1016/1353-8292(96)00004-4.

Gesler, W.M. (2003) *Healing Places*. Rowman & Littlefield, Lanham, Maryland.

Global Wellness Institute (2018) Wellness industry statistics & facts. Available at: https://globalwellnessinstitute.org/press-room/statistics-and-facts/ (accessed 9 September, 2019).

Gonzalez, L.T.V., Mariz, C.L. and Zahra, A. (2019) World youth day: Contemporaneous pilgrimage and hospitality. *Annals of Tourism Research* 76, 80–90. DOI: 10.1016/j.annals.2019.03.005.

Hadjichristodoulou, C., Mouchtouri, V., Soteriades, E.S., Vaitsi, V., Kolonia, V. *et al.* (2005) Mass gathering preparedness: the experience of the Athens 2004 Olympic and Para-Olympic Games. *Journal of Environmental Health* 67(9), 52–57.

Hahné, S.J.M., Gray, S.J., Aguilera, J.-F., Crowcroft, N.S., Nichols, T. *et al.* (2002) W135 meningococcal disease in England and Wales associated with Hajj 2000 and 2001. *The Lancet* 359(9306), 582–583. DOI: 10.1016/S0140-6736(02)07716-4.

Hall, C.M. (2011) Health and medical tourism: A kill or cure for global public health? *Tourism Review. 66(1/* 2, 4–15.

Harvey, K., Esposito, D.H., Han, P., Kozarsky, P., Freedman, D.O. *et al.* (2013) Surveillance for travel-related disease—GeoSentinel surveillance system, United States, 1997–2011. *Morbidity and Mortality Weekly Report: Surveillance Summaries* 62(3), 1–23.

Haworth, E., Barasheed, O., Memish, Z.A., Rashid, H. and Booy, R. (2013) Prevention of influenza at Hajj: applications for mass gatherings. *Journal of the Royal Society of Medicine* 106(6), 215–223. DOI: 10.1258/jrsm.2012.120170.

Henderson, J.C. (2011) Religious tourism and its management: The Hajj in Saudi Arabia. *International Journal of Tourism Research* 13(6), 541–552. DOI: 10.1002/jtr.825.

Hochedez, P., Bricaire, F., Lecso, M., Caumes, E., Valin, N. *et al.* (2009) Skin and soft tissue infections in returning travelers. *The American Journal of Tropical Medicine and Hygiene* 80(3), 431–434. DOI: 10.4269/ajtmh.2009.80.431.

Hsieh, Y.-H., Ngai, K.M., Burkle, F.M. and Hsu, E.B. (2009) Epidemiological characteristics of human stampedes. *Disaster Medicine and Public Health Preparedness* 3(4), 217–223. DOI: 10.1097/DMP.0b013e3181c5b4ba.

Huber, V. (2016) International bodies: the pilgrimage to Mecca and international health regulations. In: Tagliacozzo, E. and Toorawa, S.M. (eds) *The Hajj: Pilgrimage in Islam*. 175. Cambridge University Press, New York.

Hufford, D.J. (1985) Ste. Anne de Beaupré: Roman Catholic pilgrimage and healing. *Western Folklore* 44(3), 194–207. DOI: 10.2307/1499835.

Illiyas, F.T., Mani, S.K., Pradeepkumar, A.P. and Mohan, K. (2013) Human stampedes during religious festivals: a comparative review of mass gathering emergencies in India. *International Journal of Disaster Risk Reduction* 5, 10–18. DOI: 10.1016/j.ijdrr.2013.09.003.

Issack, M.I. and Ragavoodoo, C. (2002) Hajj-related Neisseria meningitidis serogroup W135 in Mauritius. *Emerging Infectious Diseases* 8(3), 333–334. DOI: 10.3201/eid0803.010372.

Johansson, A., Batty, M., Hayashi, K., Al Bar, O., Marcozzi, D. *et al.* (2012) Crowd and environmental management during mass gatherings. *The Lancet Infectious Diseases* 12(2), 150–156. DOI: 10.1016/S1473-3099(11)70287-0.

Keystone, J.S. (2005) Travel-related hepatitis B: risk factors and prevention using an accelerated vaccination schedule. *The American Journal of Medicine* 118(10), 63–68. DOI: 10.1016/j.amjmed.2005.07.019.

Korstanje, M.E. and Clayton, A. (2012) Tourism and terrorism: conflicts and commonalities. *Worldwide Hospitality and Tourism Themes* 4(1), 8–25. DOI: 10.1108/17554211211198552.

Kuo, H.-I., Chen, C.-C., Tseng, W.-C., Ju, L.-F. and Huang, B.-W. (2008) Assessing impacts of SARS and avian flu on international tourism demand to Asia. *Tourism Management* 29(5), 917–928. DOI: 10.1016/j.tourman.2007.10.006.

Leder, K. and Newman, D. (2005) Respiratory infections during air travel. *Internal Medicine Journal* 35(1), 50–55. DOI: 10.1111/j.1445-5994.2004.00696.x.

Löscher, T., Keystone, J.S. and Steffen, R. (1999) Vaccination of travelers against hepatitis A and B. *Journal of Travel Medicine* 6(2), 107–114. DOI: 10.1111/j.1708-8305.1999.tb00841.x.

Mangili, A. and Gendreau, M.A. (2005) Transmission of infectious diseases during commercial air travel. *The Lancet* 365(9463), 989–996. DOI: 10.1016/S0140-6736(05)71089-8.

Manoochehry, S. and Rasouli, H.R. (2017) Recurrent human tragedy during Hajj. *International Journal of Travel Medicine and Global Health* 5(1), 36–37. DOI: 10.15171/ijtmgh.2017.07.

McCarthy, M. (2016) Zika virus outbreak prompts US to issue travel alert to pregnant women. *BMJ* 352, i306. DOI: 10.1136/bmj.i306.

McKercher, B. and Chon, K. (2004) The over-reaction to SARS and the collapse of Asian tourism. *Annals of Tourism Research* 31(3), 716–719. DOI: 10.1016/j.annals.2003.11.002.

Meade, M.S. and Emch, M. (2010) *Medical Geography*, 3rd edn. The Guilford Press, London and New York.

Memish, Z.A. (2002) Meningococcal disease and travel. *Clinical Infectious Diseases* 34(1), 84–90. DOI: 10.1086/323403.

Memish, Z.A. (2010) The Hajj: communicable and non-communicable health hazards and current guidance for pilgrims. *Euro Surveillance* 15(39), 1–4.

Memish, Z.A., Stephens, G.M., Steffen, R. and Ahmed, Q.A. (2012) Emergence of medicine for mass gatherings: lessons from the Hajj. *The Lancet Infectious Diseases* 12(1), 56–65. DOI: 10.1016/S1473-3099(11)70337-1.

Memish, Z.A., Zumla, A., Alhakeem, R.F., Assiri, A., Turkestani, A. *et al.* (2014) Hajj: infectious disease surveillance and control. *The Lancet* 383(9934), 2073–2082. DOI: 10.1016/S0140-6736(14)60381-0.

Memish, Z.A. and Osoba, A.O. (2006) International travel and sexually transmitted diseases. *Travel Medicine and Infectious Disease* 4(2), 86–93. DOI: 10.1016/j.tmaid.2005.01.003.

Messenger, T. (1999) *Holy Leisure: Recreation and Religion in God's Square Mile*. University of Minnesota Press, Minnesota.

Mishra, S. (2011) *Pilgrimage, Politics, and Pestilence: The Haj from the Indian Subcontinent, 1860-1920*. Oxford University Press, New York.

Moattari, A., Emami, A., Moghadami, M. and Honarvar, B. (2012) Influenza viral infections among the Iranian Hajj pilgrims returning to Shiraz, Fars Province, Iran. *Influenza and Other Respiratory Viruses* 6(6), e77–e79. DOI: 10.1111/j.1750-2659.2012.00380.x.

Molloy, M.S. (2009) Management of mass gatherings. In: Koenig, K.L. and Schultz, C.H. (eds) *Koenig and Schultz's Disaster Medicine: Comprehensive Principles and Practices*. Cambridge University Press, Cambridge, UK, pp. 228–252.

Norman, A. and Johnson, M. (2011) World Youth Day: the creation of a modern pilgrimage event for evangelical intent. *Journal of Contemporary Religion* 26(3), 371–385. DOI: 10.1080/13537903.2011.616034.

Olapade-Olaopa, E.O., Alonge, T.O., Amanor-Boadu, S.D., Sanusi, A.A., Alese, O.B. *et al.* (2006) On-site physicians at a major sporting event in Nigeria. *Prehospital and Disaster Medicine* 21(01), 40–44. DOI: 10.1017/S1049023X00003307.

Olsen, D.H. (2006) Management issues for religious heritage attractions. In: Timothy, D.J. and Olsen, D.H. (eds) *Tourism, Religion and Spiritual Journeys*. Routledge, London and, New York, pp. 104–118.

Olsen, D.H. (2019a) Religion, spirituality, and pilgrimage in a globalising world. In: Timothy, D.J. (ed.) *Handbook of Globalisation and Tourism*. Edward Elgar, London, pp. 270–283.

Olsen, D.H. (2019b) Best practice and sacred site management: the case of Temple Square in Salt Lake City, Utah. In: Griffins, M. and Wiltshier, P. (eds) *Managing Religious Tourism*. CABI, Wallingford, UK, pp. 65–78.

Olsen, D.H. and Ron, A.S. (2013) Managing religious heritage attractions: the case of Jerusalem. In: Garrod, B. and Fyall, A. (eds) *Contemporary Cases in Heritage: Volume 1*. Goodfellow Publishers, Oxford, UK, pp. 51–78.

Olsen, D.H. and Timothy, D.J. (2018) Tourism, Salt Lake City, and the cultural heritage of Mormonism. In: Butler, R. and Suntikul, W. (eds) *Tourism and Religion: Issues, Trends and Implications*. Channel View Publications, Bristol, UK, pp. 250–269.

Ostroff, S.M. and Kozarsky, P. (1998) Emerging infectious diseases and travel medicine. *Infectious Disease Clinics of North America* 12(1), 231–241. DOI: 10.1016/S0891-5520(05)70420-7.

Palka, J.W. (2014) *Maya Pilgrimage to Ritual Landscapes: Insights from Archaeology, History, and Ethnography*. University of New Mexico Press, Albuquerque.

Petersen, E., Wilson, M.E., Touch, S., McCloskey, B., Mwaba, P. *et al.* (2016) Rapid spread of Zika Virus in the Americas -implications for public health preparedness for mass gatherings at the 2016

Brazil Olympic Games. *International Journal of Infectious Diseases* 44, 11–15. DOI: 10.1016/j.ijid.2016.02.001.

Petsalis-Diomidis, A. (2006) Amphiaraos present: images of healing pilgrimage in ancient Greece. In: Maniura, R. and Shepherd, R. (eds) *Presence: The Inherence of the Prototype within Images and Other Objects*. Routledge, Abingdon, UK, pp. 205–229.

Rafiq, S.M., Rashid, H., Haworth, E. and Booy, R. (2009) Hazards of hepatitis at the Hajj. *Travel Medicine and Infectious Disease* 7(4), 239–246. DOI: 10.1016/j.tmaid.2008.09.008.

Ratnam, I., Leder, K., Black, J. and Torresi, J. (2013) Dengue fever and international travel. *Journal of Travel Medicine* 20(6), 384–393. DOI: 10.1111/jtm.12052.

Rondy, M., Koopmans, M., Rotsaert, C., Van Loon, T., Bejaars, B. *et al.* (2011) Norovirus disease associated with excess mortality and use of statins: a retrospective cohort study of an outbreak following a pilgrimage to Lourdes. *Epidemiology and Infection* 139(3), 453–463. DOI: 10.1017/S0950268810000993.

Salamati, P. and Rahimi-Movaghar, V. (2016) Hajj Stampede in Mina, 2015: need for intervention. *Archives of Trauma Research* 5(2), e36308. DOI: 10.5812/atr.36308.

Scott, R.A. (2011) *Miracle Cures: Saints, Pilgrimage, and the Healing Powers of Belief*. University of California Press, Berkeley.

Shackley, M. (2001a) *Managing Sacred Sites: Service Provision and Visitor Experience*. Continuum, London.

Shackley, M. (2001b) Sacred world heritage sites: balancing meaning with management. *Tourism Recreation Research* 26(1), 5–10. DOI: 10.1080/02508281.2001.11081171.

Shackley, M. (2003) Management challenges for religion-based attractions. In: Fyall, A., Garrod, B. and Leask, A. (eds) *Managing Visitor Attractions: New Directions*. Butterworth-Heinemann, Oxford, UK, pp. 159–170.

Shackley, M. (2006) Empty bottles at sacred sites: Religious retailing at Ireland's national shrine. In: Timothy, D.J. and Olsen, D.H. (eds) *Tourism, Religion and Spiritual Journeys*, pp. 94–103.

Shafi, S., Booy, R., Haworth, E., Rashid, H. and Memish, Z.A. (2008) Hajj: health lessons for mass gatherings. *Journal of Infection and Public Health* 1(1), 27–32. DOI: 10.1016/j.jiph.2008.08.008.

Shafi, S., Dar, O., Khan, M., Khan, M., Azhar, E.I., McCloskey, B. *et al.* (2016) The annual Hajj pilgrimage-minimizing the risk of ill health in pilgrims from Europe and opportunity for driving the best prevention and health promotion guidelines. *International Journal of Infectious Diseases* 47, 79–82. DOI: 10.1016/j.ijid.2016.06.013.

Shinde, K.A. (2018) Palkhi: a moving sacred town. In: Olsen, D.H. and Trono, A. (eds) *Religious Pilgrimage Routes and Trails: Sustainable Development and Management*. CAB International, Wallingford, UK, pp. 150–166.

Singh, R.P.B. (2006) Pilgrimage in Hinduism: historical context and modern perspectives. In: Timothy, D.J. and Olsen, D.H. (eds) *Tourism, Religion and Spiritual Journeys*. Routledge, London and, New York, pp. 220–236.

Smith, M. and Puczkó, L. (2008) *Health and Wellness Tourism*. Routledge, London and New York.

Sokhna, C., Mboup, B.M., Sow, P.G., Camara, G., Dieng, M. *et al.* (2017) Communicable and non-communicable disease risks at the grand Magal of Touba: the largest mass gathering in Senegal. *Travel Medicine and Infectious Disease* 19, 56–60. DOI: 10.1016/j.tmaid.2017.08.005.

Solomon, M., Baum, S., Benenson, S. and Schwartz, E. (2011) Tropical skin infections among Israeli travelers. *The American Journal of Tropical Medicine and Hygiene* 85(5), 868–872. DOI: 10.4269/ajtmh.2011.10-0471.

Spencer, B., Steele, W., Custer, B., Kleinman, S., Cable, R. *et al.* (2009) Risk for malaria in United States donors deferred for travel to malaria-endemic areas. *Transfusion* 49(11), 2335–2345. DOI: 10.1111/j.1537-2995.2009.02290.x.

St. John, R.K., King, A., de Jong, D., Bodie-Collins, M., Squires, S.G. *et al.* (2005) Border screening for SARS. *Emerging Infectious Diseases* 11(1), 6–10. DOI: 10.3201/eid1101.040835.

Stausberg, M. (2011) *Religion and Tourism: Crossroads, Destinations, and Encounters*. Routledge, New York.

Steffen, R. (2010) The risk of meningococcal disease in travelers and current recommendations for prevention. *Journal of Travel Medicine* 17(suppl 1), 9–17. DOI: 10.1111/j.1708-8305.2010.00449.x.

Steffen, R., Bouchama, A., Johansson, A., Dvorak, J., Isla, N. *et al.* (2012) Non-communicable health risks during mass gatherings. *The Lancet Infectious Diseases* 12(2), 142–149. DOI: 10.1016/S1473-3099(11)70293-6.

Steffen, R. (2017a) Travel health -travel medicine. *Journal of Health Specialties* 5(1), 42–44. DOI: 10.4103/2468-6360.198796.

Steffen, R. (2017b) Epidemiology of travellers' diarrhea. *Journal of Travel Medicine* 24(suppl_1), S2–S5. DOI: 10.1093/jtm/taw072.

Taha, M.-K., Achtman, M., Alonso, J.-M., Greenwood, B., Ramsay, M. *et al.* (2000) Serogroup W135 meningococcal disease in hajj pilgrims. *The Lancet* 356(9248), 2159. DOI: 10.1016/S0140-6736(00)03502-9.

Tatem, A.J., Huang, Z., Das, A., Qi, Q., Roth, J. *et al.* (2012) Air travel and vector-borne disease movement. *Parasitology* 139(14), 1816–1830. DOI: 10.1017/S0031182012000352.

Tayan, O., Al BinAli, A.M. and Kabir, M.N. (2014) Analytical and computer modelling of transportation systems for traffic bottleneck resolution: a Hajj case study. *Arabian Journal for Science and Engineering* 39(10), 7013–7037. DOI: 10.1007/s13369-014-1231-3.

Timothy, D.J. and Iverson, T. (2006) Tourism and Islam: considerations of culture and duty. In: Timothy, D.J. and Olsen, D.H. (eds) *Tourism, Religion and Spiritual Journeys*. Routledge, London and New York, pp. 202–221.

Timothy, D.J. and Olsen, D.H. (2006) *Tourism, Religion andSpiritual Journeys*. Routledge, London and New York.

UNWTO (2019) *UNWTO Tourism Highlights, 2019 Edition*. UNWTO, Madrid. Available at: https://www.e-unwto.org/doi/pdf/10.18111/9789284421152 (accessed 9 September 2019).

Van Herck, K., Castelli, F., Zuckerman, J., Nothdurft, H., Van Damme, P. *et al.* (2004) Knowledge, attitudes and practices in travel-related infectious diseases: the European Airport survey. *Journal of Travel Medicine* 11(1), 3–8. DOI: 10.2310/7060.2004.13609.

Voigt, C. and Pforr, C. (eds) (2013) *Wellness Tourism: A Destination Perspective*. Routledge, London and New York.

Weir, E. (2002) Travel warning: eosinophilic meningitis caused by rat lungworm. *Canadian Medical Association Journal* 166(9), 1184–1184.

Wetmore Jr, K.J. (ed.) (2017) *The Oberammergau Passion Play: Essays on the 2010 Performance and the Centuries-Long Tradition*. McFarland, Jefferson, North Carolina.

Wilder-Smith, A. (2006) The severe acute respiratory syndrome: impact on travel and tourism. *Travel Medicine and Infectious Disease* 4(2), 53–60. DOI: 10.1016/j.tmaid.2005.04.004.

Wilder-Smith, A. (2017) Closing the gap in travel medicine. *Journal of Travel Medicine* 24(4), 1–3. DOI: 10.1093/jtm/tax027.

Woodward, S.C. (2004) Faith and tourism: planning tourism in relation to places of worship. *Tourism and Hospitality Planning & Development* 1(2), 173–186. DOI: 10.1080/1479053042000251089.

WTTC (2015) Tourism and the spread of disease. Available at: https://medium.com/@WTTC/tourism-and-the-spread-of-disease-cc73e5ed3471 (accessed 16 April, 2019).

Yamin, M. and Albugami, M.A. (2014) An architecture for improving Hajj management. In: Liu, K., Gulliver, S.R., Li, W. and Yu, C. (eds) *Service Science and Knowledge Innovation*. Springer, Berlin and Heidelberg, pp. 187–196.

Yamin, M. (2015) Secure and healthy Hajj management: a technological overview. *American Academic & Scholarly Research Journal* 7(3), 195–202.

Yassen, M.A. and Sameer, M.A. (2006) Emergency room to the intensive care unit in Hajj. The chain of life. *Saudi Medical Journal* 27(7), 937–941.

Yezli, S., Assiri, A.M., Alhakeem, R.F., Turkistani, A.M. and Alotaibi, B. (2016) Meningococcal disease during the Hajj and Umrah mass gatherings. *International Journal of Infectious Diseases* 47, 60–64. DOI: 10.1016/j.ijid.2016.04.007.

Yu, V.L. and Madoff, L.C. (2004) ProMED-mail: an early warning system for emerging diseases. *Clinical Infectious Diseases* 39(2), 227–232. DOI: 10.1086/422003.

Zeller, H. (1876) *Wild Flowers of the Holy Land*. J. Nisbet, London.

Zeng, B., Carter, R.W. and De Lacy, T. (2005) Short-term perturbations and tourism effects: the case of SARS in China. *Current Issues in Tourism* 8(4), 306–322. DOI: 10.1080/13683500508668220.

Zuckerman, J.N. (2002) Travel medicine. *BMJ* 325(7358), 260–264.

10 The Natural Environment and Waste Management at the *Hajj*

Jahanzeeb Qurashi*
University of Wales Trinity Saint David, UK

Introduction

Since the beginnings of humanity, people have travelled to distant lands and places for economic and political reasons. From Old Testament times to the Roman Empire, people enjoyed forms of resort-based tourism not dissimilar to present-day mass tourism to beaches and inclusive resorts (Sigaux, 1966; Young, 1973; Casson, 1974). For many people, pilgrimage to sites deemed sacred was a great motivator for travel. Early Neolithic temples at Malta, Turkey, and Britain stand as evidence of this (Adkins and Adkins, 1985; Grima, 2001; Rountree, 2002). Prior to the advent of modern transportation and information technologies, these pilgrimages were fraught with danger and hardship, with the adversity being seen as a part of the penitential nature of pilgrimage (Jebb, 1986). During the medieval period, pilgrimage was a widespread social institution practised within the world's principal religious traditions of Christianity, Islam, Hinduism, and Buddhism (Digance, 2006).

Travel for religious purposes is still an important motivator today, with the number of people travelling for this reason increasing after World War II. This is the case in Islam, where travel to Makkah for the annual *Hajj* pilgrimage is the fifth mandatory pillar of the faith – the other pillars being the faith of belief (*Shahada*), praying five times daily (*Salah*), giving to the poor (*Zakat*), and fasting during Ramadan (*Sawm*) – and is a requirement for those who can afford it and are physically able to go (Haq and Jackson, 2009). As Allah (God) has said in the holy book of the Quran: ' ... proclaims to mankind the pilgrimage [Hajj]; they will come to you on foot and on every lean camel; they will come from every distant pass' (Surah Al-Hajj, 22, p. 27). Every year, between 2–3 million Muslims make the pilgrimage to Makkah (United Press International, 2010). Only Muslims can enter the city of Makkah to perform the *Hajj* and go into the Grand Mosque (Masjid al-Haram) that houses the *Kaaba* (also known as 'The Cube' or 'Sacred House').

Makkah itself, with more than 2 million inhabitants, is the capital city of the Kingdom of Saudi Arabia (KSA). The holy city's importance far outweighs its capital status, as it serves as the capital of Islam because of its hosting of the annual *Hajj* pilgrimage for more than 1430 years (Gwyn, 1989). The city came about following God's instructions to Ibrahim (PBUH) to build a monument at the site of a spring known as the *Kaaba*. While worshippers from many faiths travelled to worship at the site, in AD 630 the Prophet Mohammed (PBUH) led a group of Muslims there in the first official *Hajj*. During the course of this pilgrimage, Mohammed (PBUH) and his

*j.qurashi@uwtsd.ac.uk

followers destroyed religious idols placed there by polytheistic worshippers and re-dedicated the site in the name of Allah (God), (Qurashi, 2017). Today, Makkah occupies an area of 33 km^2, and is further divided into three areas: Mina (8 km^2), Arafat (13 km^2), and Muzdalifah (12 km^2) (Makkah Municipality, 2010). During the *Hajj*, pilgrims spend their first day in the tent city of Mina, the next day 12 hours in the field of Arafat, a night in Muzdalifah, and then return to the tent city of Mina for four to five days.

Over time the number of pilgrims participating in the *Hajj* has been dramatically increasing. In 1950, approximately 250,000 pilgrims participated. This number increased to 3.16 million in 2012, and in 2015 there was a slight decrease to 2.95 million (Mintel, 2016). Moreover, more than 12 million pilgrims visit the holy cities of Makkah and Madinah each year (Arab News, 2013), with more than 17 million pilgrims expected to visit these cities by 2025 (Wainwright, 2015). More broadly, international religious tourism to KSA increased by more than 2% annually between 2006 and 2010; in 2010 alone, almost 4.9 million international religious pilgrims travelled to KSA, representing 17.1% of worldwide religious tourist arrivals (Jafari and Scott, 2010).

This increase in pilgrims and religious tourists has led to concerns regarding the environmental impact of the visitors, particularly during the *Hajj* pilgrimage. Much has been written on the environmental issues stemming from the *Hajj*, such as solid waste management in Mina (Alsebaei, 2014), the role of Saudi waste management and recycling plants during the *Hajj* and Umrah (Anjum *et al.*, 2016), and the management of landfills and air pollution in Makkah during the *Hajj* (Habeebullah, , 2013). However, this chapter focuses on examining how the pilgrimage, hospitality, and tourism sectors are threatening the environment of the *Hajj*, how the KSA authorities are managing waste during the *Hajj*, and what the government's future plans are for ensuring the sustainability of the environment during future *Hajj* events.

To accomplish this, the author engaged in one-on-one and face-to-face semi-structured interviews, supplemented by participant observation, during the 2016 *Hajj*. The selection of participants' interviews was undertaken through personal connections as well as random and snowballing sampling strategies with pilgrims, hoteliers, and Makkah municipality workers. In total, 78 people were interviewed, among which were ten hoteliers from Makkah, eight others were Makkah municipality workers, and the rest were pilgrims from different areas of the world. After discussing the relationship between tourism, pilgrimage, and the environment, the chapter focuses on the types of environmental problems pilgrims and the hospitality and tourism sectors create during the *Hajj*. The chapter then turns to examine how the Saudi government is attempting to cope with these environmental challenges as well as slaughterhouse waste, before concluding.

Tourism, Pilgrimage, and Environment

As noted above, pilgrimage is one of the oldest and most famous forms of expressing religious worship. Pilgrimage has also been considered one of the oldest forms of tourism and is an integral part of the tourism industry (Rinschede, 1990; Shinde, 2008). Pilgrimage is not only a symbol of Christian life with deep roots in the Bible, but is also outlined in the holy books of other religions, such as the Torah for Jews and the Quran for Muslims. However, scholars of religious tourism and pilgrimage have only begun to scratch the surface of both the historical and the present-day economic, sociocultural, and environmental effects of religiously motivated travel (e.g. Shackley, 2001; Olsen and Timothy, 2006; Pourtaheri *et al.*, 2012; Vijayanand, 2012; Qurashi, 2017). Of these three areas of scholarship, the effects of pilgrimage and religious tourism on the natural and built environment are lacking (Shinde, 2007a). While writing on the interface between environmental and religious tourism and pilgrimage is growing (e.g. Nelson, 1998; Shinde, 2007b; El Hanandeh, 2013; Alipour *et al.*, 2017), much of the research is hypothetical and theory-based.

The various environments at pilgrimage locations affected by the arrival of pilgrims are numerous and characteristically revolve around short or long-pilgrim term needs and requirements such as accommodation, food and beverages, and other tourism infrastructure facilities.

These needs generate enormous amounts of different kinds of waste (e.g. effluent, water, food) as well as air pollution. These environmental effects occur due to overtourism – or overpilgrimage in this case; the nature of activities undertaken by pilgrims; and the commodification of religious destinations and the pilgrimage journeys.

Unfortunately, while different religious theologies point to the importance of environmental stewardship, the issues of climate change and conservation, while a concern for many religious site managers, are not always a high priority for pilgrims and tourists (Sati, 2015; Luthy, 2016). Indeed, while much of the research on environmental sustainability and conservation in the context of pilgrimage and religious tourism is on western religious sites and similar sites in India, environmental issues related to the *Hajj* are just becoming a point of interest among scholars (e.g. Al-Jeelani, 2009; El Hanandeh, 2013; Alsebaei, 2014). As such, the focus of the rest of the chapter is to explore the environmental issues of the *Hajj* as created by pilgrims and the tourism and hospitality industry, and how the KSA government is attempting to solve these environmental issues.

Environmental Issues at the *Hajj*: Pilgrims

The city of Makkah is bordered by a sacred area called Haram in which pilgrims accomplish the diverse, multifaceted rites of the *Hajj*. It is here where the circumambulation of the *Kaaba* (black cube), running and walking between the hills of Al-Safa and Al-Marwah inside the grand mosque, staying for a day in the field of Arafat, spending a night in the rocky field of Muzdalifah to collect pebbles to stone the three devils in Mina, and staying in the tent city of Mina, takes place (Robinson, 1999). During fieldwork in this area, the author observed that pilgrims created an enormous amount of waste, particularly at Mina and Arafat (where they spent most of their time being sedentary), both in terms of sewage through the excessive use of water or in terms of solid waste (e.g. soft drinks cans, paper, and plastic). Indeed, pilgrims throw their waste on to the street while they are walking, finding that the waste bins at the corner of the street or tents are too far away for their convenience. Also, many camp janitors dispose of camp waste in the streets, particularly when the available waste bins are full (Fig. 10.1). Also, at the exit of Devil's Bridge, pilgrims sit on the road and shave their hair to fulfil one of the requirements of *Hajj* rites (Fig. 10.2), which activity is both hygienically and environmentally concerning.

These types of waste quickly become unsustainable and an important health issue in the context of 3 million pilgrims. For example, Alsebaei (2014) suggests that if the waste generated in the tent city of Mina is assumed to be the same as the quantity of waste disposed of in Makkah landfill, then on average pilgrims generate 3410.4 tons of solid waste a day (0.71 $kg.m^{-2}$ of waste per day). This would mean that the waste generation rate per pilgrim is 4.62 kg each day. Moreover, the City of Makkah itself produces 2.4,000 tons of municipal solid waste every day. While this solid waste amount increases to between 3.1 and 4.6,000 tons per day during Ramadan (the month of fasting) and the *Hajj* (Nizami *et al.*, 2015), the highest waste generation occurs during *Zulhijjah* (the month of Pilgrimage) and the last ten days of Ramadan.

When asked about their views on environmental sustainability and waste disposal practices, pilgrims seemed indifferent, and demonstrated very little connection with the environmental issues of the *Hajj* and its conservation. The majority of pilgrims interviewed stated that it is the duty of the KSA authorities to deal with the environmental issues, not the pilgrims. This view was repeated by pilgrims of all socioeconomic conditions – by pilgrims from developing countries to affluent pilgrims from Saudi Arabia. These environmental perspectives go directly against Islamic belief, which is implicitly conservationist. The Quran, for example, includes verses that support the idea of environmental stewardship and sustainability: 'Eat and drink from the provision of Allah, and do not commit abuse on the earth, spreading corruption' (Quran 2:60). Considering Islam's stance on environmental conservation, these views by pilgrims seem confusing at best.

Fig. 10.1. Waste dumped by pilgrims and camp janitors during the 2016 *Hajj* (author).

Fig. 10.2. Pilgrims cutting their hair at the Devil's Bridge during the 2016 *Hajj* (author).

Environmental Issues at the *Hajj*: The Hospitality and Tourism Industry

With increasing numbers of pilgrims participating in the *Hajj* since 1950, the city of Makkah has changed greatly in terms of its urban and pilgrimage infrastructure. Even today, continued development in the areas of rapid and extensive transportation and accommodation facilities, and telecommunication infrastructure, continues apace. This has contributed to the growth of the pilgrimage and religious tourism sector, which can be demonstrated through reviewing the accommodation sector in Makkah. Colliers (2017) reports that during the second quarter of 2017, Makkah enjoyed a significant increase on the second quarter of 2016 in both occupancy and average daily rates, the latter reaching US$600 to US$700 in the branded hotel market during the *Hajj*. In the month of June 2016 alone, the market achieved an occupancy of 80.7%, representing a 24% increase from the same month in 2015. Similarly, during the Holy Month of Ramadan, the average daily rate reached US$467, representing a 70% year-on-year increase from 2015. The holy city of Madinah also experienced an increase in average daily rates and occupancy levels, though not as significant as those in Makkah (Colliers, 2017). Because of increasing room demand, Colliers (2017) forecasts that by the end of 2019 there will be 39,898 hotel rooms in Makkah, a large increase over 21,209 rooms in the city at the end of 2017, due in large part to the building of new hotels in the city (Ascoura, 2013).

In addition, owing to the increase in the number of pilgrims visiting Makkah and the high summer temperatures, the KSA is planning to expand the grand Mosque Al-Haram and install air conditioning in the mosque (Henderson, 2011). Redfern (2010, p. 32) notes that current capacity for the grand Mosque Al-Haram is 700,000 pilgrims, but following expansion the capacity of the mosque will be raised to 2.2 million. Further extensions to the mass transit system will link to the Grand Mosque of Al Haram with a further four-mast transit lines under construction, including a line from Jeddah Airport to Makkah (Khalid, 2016).

With this economic and infrastructure expansion comes the potential for increased environmental issues. Bed and breakfasts and star-ranked hotels contribute to solid waste management (SWM) problems. One question asked of star-ranked hotel managers in Makkah dealt with whether these internationally branded hotel chains had developed green policies regarding decreasing their environmental impact on Makkah. The majority of interviewees said their hotels did not have any green policies; those hotels that had a green policy did not follow the policy in practice. Hotel managers noted that food waste does triple in amount during Ramadan and *Hajj* thanks to lavish hotel buffets and a general pilgrim habit of over-consumption. Additionally, a hotel manager told the author that in Makkah, much of the solid waste generated by hotels goes to the local landfill, including plastic water bottles, cups, styrofoam take-out containers with half-eaten food, and plastic carrier bags. Also, the old buses that transport tourists and pilgrims fill the air with carbon dioxide fumes throughout the urban area around the hotels (Bhattay, 2017). When this air pollution reaches certain levels, the emitted pollutants can lead to adverse effects on pilgrims, including contributing to respiratory disease, asthma, cardiovascular disease and lung cancer, and even death (Habeebullah, , 2013; Alsebaei, 2014; Khwaja, 2017; see also Chapter 9, this volume). In addition to these local environmental issues, from a broader perspective, transportation taken by pilgrims from around the world, such as planes, boats, and cars, to participate in the *Hajj*, also adds to global pollution rates.

Environmental Issues at the *Hajj*: KSA Authorities and SWM

Within the KSA, the Ministry of Municipal and Rural Affairs (MMRA) is responsible for creating the policies regarding environmental sustainability and the *Hajj* while the Makkah municipal council is responsible for the implementation of those policies. The MMRA has proposed and undertaken many projects related to the protection and conservation of the urban and natural environment during the *Hajj*, including crowd management, increasing carrying capacity, providing pilgrims with improved health care, and the introduction of certain technologies to better manage these issues throughout the *Hajj*. However, one problem that has not been fully considered and solved is the problem of SWM during the *Hajj*. Because of the lack of resources, Makkah municipal authorities have tried to minimize the impacts of this problem by making the waste less visible.

Of the waste produced during the *Hajj*, 40% is comprised of organic waste, which comes from different sources such as hotels, restaurants, canteens, and local residences, in addition to pilgrims (Adhikari *et al.*, 2008). Plastics are also a major solid waste management concern.

During the holy month of Ramadan and the *Hajj* in particular, great amounts of plastic waste are produced, including plastic water bottles, bags, food wrappers, and utensils (Alruqaie, 2012). Only 15–20% of all produced plastic waste is sorted and recycled, with the rest being thrown into the local landfill.

The present management arrangement of SWM in Saudi Arabia mostly involves gathering and discarding waste in landfill sites (Ouda *et al.*, 2013). While the amount of waste put into Makkah landfill is about 1,800–2,000 tons a day during non-*Hajj* times, during the *Hajj* the quantity of waste put into landfill is between 3000 to 4000 tons. The landfill facility, situated at Buraiman, Jeddah (a neighbouring city to Makkah), receives about 1.5 million tons of non-*Hajj* solid waste every year (Zafar, 2015), but 4500 tons a day during the *Hajj* (Ouda *et al.*, 2013).

While the KSA attempts to improve solid waste management policies and practices, on occasion unforeseen problems arise. For example, between 2010 and 2016, the KSA had a lack of sanitation workers to help remove the solid waste created by pilgrims. This is because the Saudi work visa policy is not only complicated at best, but is also prone to corruption. As a case in point, in 2010, 2121 out of the 4000 visas allotted to external sanitation workers were illegally sold to other people (Garsan, 2011). One interviewee told the author that they felt that the Makkah municipality approach to solid waste management was 'irrational'. For instance, the interviewee stated that there are no checks and balances regarding the gathering of solid waste. The interviewee gave the example of where solid waste removal is not equally done inside the different Mina pilgrim camps, as appropriate waste removal depended on the attitude and behaviour of the camp managers as well as the sanitation workers assigned to those camps. Although any camp manager who disposes of camp waste on the street is fined 10,000 Saudi Riyal (approximately US$2666), many camp managers choose to not follow waste management regulations and throw their waste outside the waste bins and on to the street because of the lack of government enforcement (Fig. 10.3). Indeed, the author noted that in one instance, rather than clean accumulated waste, sanitation workers began to ask pilgrims for charity money and ignored their assigned duties. Additionally, the author observed that many sanitation workers,

Fig. 10.3. Garbage covering a compactor box (author).

when picking through the solid waste, focused on searching for valuables that pilgrims might have lost in the shape of money, mobile phones, and jewellery rather than recyclable waste. Other interviewees noted additional problems with the current waste management system. For example, many of the waste bins and compactor boxes are emptied on an inconsistent basis, are often damaged, leak dirty water and produce filthy odours. As such, while waste management policies may be sound, the problem seems to be lack of implementation of and accountability regarding these policies.

Slaughterhouse Waste During the *Hajj*

Slaughtering a healthy four-legged Halal animal is mandatory during *Hajj* by all pilgrims. To fulfil the requirements of this ritual, a slaughterhouse operates in Mina. During the 2016 *Hajj*, the minimum amount paid by each of the pilgrims for an animal was 550 Saudi Riyal ($146). The slaughterhouse is divided into two sections, with one section being where the animals are butchered and the other section where the animal meat is treated. A third zone is the meat supply zone where pilgrims can go to get their part of the meat.

The slaughterhouse is 50 acres in size because of the approximately 750,000 animals that need to be housed before the annual sacrifice, which occurs over the course of four days. Camels, cows, sheep, and goats are imported from other Arab states and, if there is a shortage, from other countries. Before entering the slaughterhouse, veterinarians and Sharia scholars screen each animal to ensure they are compliant with the religious standards of sacrificial animals. Meat that is not used for ritual sacrifice is distributed among the needy in the area and to other charity organizations in the KSA and other areas of the world (Brooke, 1987). This slaughterhouse remains operational for various other commercial or religious ceremonies throughout the year as required.

There have been many attempts to limit the environmental impact of the slaughterhouse. For example, much of the excess meat is canned to eliminate plastic waste. In addition, a waste biorefinery takes the animal biological waste and transforms it into energy and fertiliser (Nizami *et al.*, 2015). However according to a plant manager, the capacity of the plant generally operates at less than 10%, except for during the *Hajj*, where during the slaughter days animal biological and organic waste reaches approximately 100,000 tons, equalling approximately 1900 tons a day or 600 tons in an eight-hour stretch. The problem is that the plant's capacity is only 5.5 tons per hour. However, one of the advantages of this plant is that waste that would normally make it to the landfill is diverted into the creation of energy.

Regarding the blood produced by these animal sacrifices, large trucks carry it for disposal in an unknown location. However, Aziz *et al.* (2015) suggest that wherever this blood is dumped, this exercise is anticipated to contaminate the groundwater, and as such the blood needs to be treated prior to dumping. Hassan *et al.* (2015) also note that the air at a slaughterhouse contains methane and carbon dioxide, along with other airborne toxins, while the total suspended particulates in the air are well beyond permissible levels set by the World Health Organization (WHO).

Many private companies are contracted to take animal waste to treatment plants throughout the KSA. Upon asking to a few private drivers for these companies regarding how their company deals with this waste, the author was told that the disposal of the animal waste is a very lucrative business. Animal skins are sold to garment and shoe companies, and other biological parts sell to sports industry manufacturers to create different sports equipment such as badminton rackets. Moreover, the rest of the animal carcasses and organs are used to manufacture chicken feed.

Greening the *Hajj*

Introducing the concept of green *Hajj* is a major challenge, not only for the KSA authorities but for other stakeholders as well. All stakeholders must realise there is a social and religious duty to maintain the environmental decorum of *Hajj*. The first international effort to 'green' the *Hajj* came from the Alliance of Religions and Conservation event in November 2011 in Assisi, Italy. From this conference came about a *Green Hajj Guide* to encourage pilgrims, the tourism and hospitality industry, and the KSA authorities to reduce the impact of the *Hajj* on the Earth (Khaleafa, 2012). According to Khaleafa (2012), the *Guide* was created to suggest guidelines on how to help pilgrims and stakeholders have a greener attitude and engage in greener practices towards the *Hajj*. The *Guide* is separated into several sections, including the importance of environmental conservation and sustainability in Islamic theology; making faith-consistent purchases through buying environmentally friendly (read: no plastics) and fair trade products; limiting one's carbon footprint before, during, and after the *Hajj*; and recommendations for stakeholders responsible for the *Hajj* regarding how to make the entire event experience greener.

The KSA authorities have taken several steps towards greening the *Hajj*. For example, the government has built a plasma-chemical reactor with plasma torches to destroy waste in an environmentally friendly way (Galaly and Oost, 2017). The authorities have also begun to set quotas for pilgrims in terms of the total number allowed to perform the *Hajj* as well as only allowing pilgrims to perform the *Hajj* once every five years. The government has also invested in the Haramain High Speed Railway (finished in 2018), which connects Makkah and Medina. The Al Mashaaer Al Mugaddassah Metro line in Makkah, opened in 2010, was built to shuttle pilgrims between Makkah, Mount Arafat, Mina, and Muzdalifah. The Makkah Metro, which only operates during the *Hajj*, also serves to move pilgrims between sacred sites. The Makkah Metro is critical in helping ease the heavy congestion during *Hajj*, as it moves crowds of approximately 90,000 people per hour (Raouf, 2016). These transportation developments have brought significant change in green *Hajj* sustainability, with using public transportations and trains helping to decrease air pollution and traffic congestion.

In another effort to green the *Hajj*, the Saudi authorities have partnered with US biotechnology company Bio Wish Technologies to help prevent waste-related aromas through spraying a mixture of naturally occurring microorganisms in garbage bins and compactor bins. This programme, first started in 2014, has not only limited garbage-related odours, but has also decreased the number of insects around the garbage bins and instances of the *Hajj* cough and insect-borne infections (Bio Wish Technologies, 2016). One additional greening effort has been to construct several solar power plants that by 2030 will produce 40 gigawatts (Bellini, 2019).

Conclusion

The number of Muslims participating in the *Hajj* has increased in recent decades with improvements in modern transportation systems and the development of new, modern facilities to meet the wants and needs of pilgrims. However, as the scale of the *Hajj* has increased, so too have the environmental costs of this religious journey. As noted above, these environmental issues stem from too many pilgrims, the types of activities (physical and spiritual) undertaken by pilgrims, and the commodification of pilgrimage activities in Makkah. In the case of the *Hajj*, government officials have created national policies relating to greening the *Hajj* as well as the local Makkah environment. Some of these policies and practices include the building of advanced integrated waste management plants, changes in energy reliance on fossil fuels with the building of solar plants, and also creating renewable energy sources from waste sources.

However, these attempts to green the *Hajj* may ultimately be fruitless as long as there is a disconnect between Islamic theology related to environmental stewardship and conservation and the poor environmental actions of pilgrims and tourism and hospitality managers. Indeed, the animal sacrifices required during the *Hajj* and the need for constant potable water and food resources for all pilgrims influence the link between religious discourses and the environment. Also, while limiting the growth of the *Hajj* is a good start, environmental sustainability can only be achieved if religious institutions which are directly engaged in the *Hajj* take some of the environmental responsibility for this greening, such as educating potential pilgrims regarding the environmental issues that occur during the *Hajj*. In this way, a more environmentally mindful pilgrim can help decrease their environmental footprint and help be a part of the true spirit of the *Hajj*.

References

Adhikari, B.K., Barrington, S., Martinez, J. and King, S. (2008) Characterization of food waste and bulking agents for composting. *Waste Management* 28(5), 795–804. DOI: 10.1016/j.wasman.2007.08.018.

Adkins, L. and Adkins, R.O.Y.A. (1985) Neolithic axes from Roman sites in Britain. *Oxford Journal of Archaeology* 4(1), 69–76. DOI: 10.1111/j.1468-0092.1985.tb00231.x.

Al-Jeelani, H.A. (2009) Evaluation of air quality in the Holy Makkah during Hajj season. *Journal of Applied Sciences Research* 5(1), 115–121.

Alipour, H., Olya, H.G.T. and Forouzan, I. (2017) Environmental impacts of mass religious tourism: from residents' perspectives. *Tourism Analysis* 22(2), 167–183. DOI: 10.3727/108354217X14888192562285.

Alruqaie, I. (2012) Environmental advantage assessment of recycling food waste in Riyadh, Saudi Arabia. *Research Journal of Environmental Sciences* 6(6), 230–237. DOI: 10.3923/rjes.2012.230.237.

Alsebaei, F.A. (2014) Solid waste management and recycling during Hajj pilgrimage in Mina. PhD dissertation. Umm Al-Qura University, Kingdom of Saudi Arabia.

Anjum, M., Waqas, M., Ijaz, A. and Barkat, A. (2016) Solid waste management in Saudi Arabia: a review. *Journal of Applied Agriculture and Biotechnology* 1(1), 13–26.

Arab News (2013) Revenue from pilgrims makes 3% of Saudi GDP. Arab News. Available at: http://www.arabnews.com/revenue-pilgrims-makes-3-saudi-gdp (accessed 12 December 2016).

Ascoura, I.E. (2013) Impact of pilgrimage (Hajj) on the urban growth of the Mecca. *Journal of Educational and Social Research* 3(2), 255–263. DOI: 10.5901/jesr.2013.v3n2p255.

Aziz, A.H., Isa, H.M., Kadir, O.A., Nordin, M.N., Daud, W.R.W.A. *et al.* (2015) Study on the baseline data of solid waste management in the Holy city of Mekkah during Hajj 1427. Available at: http://www.academia.edu/8563835/STUDY_on_the_baseline_data_of_solid_waste_management_in_theology_city_of_Makkah_during_Hajj_1427 (accessed 8 November 2019).

Bellini, E. (2019) Saudi Arabia to tender 2.22 GW of solar in 2019 and wants 40 GW in 2030. PV-Magazine. Available at: https://www.pv-magazine.com/2019/01/10/saudi-arabia-to-tender-2-22-gw-of-solar-in-2019-and-wants-40-gw-in-2030/ (accessed 8 November 2019).

Bhattay, A. (2017) Hajj: how to make this three million strong Muslims pilgrimage environmentally friendly. The Independent. Available at: https://www.independent.co.uk/travel/asia/hajj-pilgrimage-green-guide-mecca-medina-muslims-environmental-friendly-travel-islam-sustainable-a7930676.html (accessed 8 November 2019).

Bio Wish Technologies (2016) Garbage and its odor. Biowishtechnologies.com. Available at: https://biowishtechnologies.com/environmental-hazards-garbage-odor/ (accessed 8 November 2019).

Brooke, C. (1987) Sacred slaughter: The sacrificing of animals at the *Hajj* and *Id al-Adha*. *Journal of Cultural Geography* 7(2), 67–88. DOI: 10.1080/08873638709478508.

Casson, L. (1974) *Travels in the Ancient World*. George Allen and Unwin, London.

Colliers (2017) Saudi Arabia Q2 2017 review: 5 key cities. Available at: https://www.colliers.com/-/media/files/emea/uae/case-studies/2017-overview/ksa-hotels-review-q2-2017.pdf (accessed 8 November 2019).

Digance, J. (2006) Religious and secular pilgrimage: journeys redolent with meaning. In: Timothy, D.J. and Olsen, D.H. (eds) *Tourism, Religion and Spiritual Journeys*. Routledge, London and New York, pp. 36–48.

El Hanandeh, A. (2013) Quantifying the carbon footprint of religious tourism: the case of Hajj. *Journal of Cleaner Production* 52, 53–60. DOI: 10.1016/j.jclepro.2013.03.009.

Galaly, R. and Oost, G, V. (2017) Environmental and economic vision of plasma treatment of waste in Makkah. *Plasma Science and Technology* 19(1), 1–10.

Garsan, A. (2011) Investigation leads to Indictment to cleaning company in Makkah. *Okaz*, 16 March. p.10.

Grima, R. (2001) An iconography of insularity: a cosmological interpretation of some images and spaces in the Late Neolithic temples of Malta. *Papers from the Institute of Archaeology* 12, 48–65. DOI: 10.5334/pia.164.

Gwyn, R. (1989) The centrality of Islam: space, form and process. *GeoJournal* 18(4), 351–359.

Habeebullah,M.T. (2013) An Analysis of Air Pollution in Makkah - a View Point of Source Identification. *Environment Asia* 6(2), 11–17.

Haq, F. and Jackson, J. (2009) Spiritual journey to Hajj: Australian and Pakistani experience and expectations. *Journal of Management, Spirituality & Religion* 6(2), 141–156.

Hassan, I.A., Basahi, J.M., Ismail, I.M., Alghamdy, M.I., Khoder, M.I. *et al.* (2015) Emissions of greenhouses gases and total suspended particulates (TSP) from slaughter house during Hajj season in Makkah, KSA. Available at: http://www.kau.edu.sa/GetFile.aspx?id=266099&fn=Dr.Iqbal-English%20version.pdf (accessed 8 November 2019).

Henderson, J.C. (2011) Religious tourism and its management: the Hajj in Saudi Arabia. *International Journal of Tourism Research* 13(6), 541–552. DOI: 10.1002/jtr.825.

Jafari, J. and Scott, N. (eds) (2010) *Tourism in the Muslim World*. Emerald, Bingley, UK.

Jebb, M. (1986) *Walkers*, 1st edn. Constable, London.

Khaleafa, A. (2012) The green Hajj guide. Available at: http://www.khaleafa.com/khaleafacom/the-green-hajj-guide (accessed 8 November 2019).

Khalid, M. (2016) Makkah is world's most profitable hotel market. Available at: http://www.khaleejtimes.com/business/economy/makkah-is-worlds-most-profitable-hotel-market (accessed 2 June 2018).

Khwaja, H.A. (2017) An assessment of air quality in the surrounding Holy places of Mecca, Saudi Arabia during hajj. ConferenceSeries.com. Available at: https://environmentalchemistry.conferenceseries.com/abstract/2017/an-assessment-of-air-quality-in-the-surrounding-holy-places-of-mecca-saudi-arabia-during-hajj (accessed 8 November 2019).

Luthy, T. (2016) "Few people know that Krishna was the first environmentalist": Religiously motivated conservation as a response to pilgrimage pressures in Vrindavan, India. In: Mostafanezhad, M., Norum, R., Shelton, E.J. and Thompson-Carr, A. (eds) *Political Ecology of Tourism: Community, Power and the Environment*. Routledge, London and New York, pp. 92–107.

Makkah Municipality (2010) Makkah City statistics. Available at: http://www.holymakkah.gov.sa (accessed 10 June 2019).

Mintel (2016) *Religious and Pilgrimage Tourism*. Mintel International Group Ltd, London.

Nelson, L.E. (1998) *Purifying the Earthly Body of God*. State University of New York Press, Albany.

Nizami, A.-S., Rehan, M., Ismail, I., Almeelbi, T. and Ouda, O. (2015) Waste biorefinery in Makkah: a solution to convert waste produced during hajj and Umrah seasons into wealth. Presented at the 15th scientific symposium for Hajj, Umrah and Madinah visit. Available at: https://www.researchgate.net/publication/277332302_Waste_Biorefinery_in_Makkah_A_Solution_to_Convert_Waste_produced_during_Hajj_and_Umrah_Seasons_into_Wealth (accessed 8 November 2019).

Olsen, D.H. and Timothy, D.J. (2006) Tourism and religious journeys. In: Timothy, D.J. and Olsen, D.H. (eds) *Tourism, Religion and Spiritual Journeys*. Routledge, London and New York, pp. 1–21.

Ouda, O.K.M., Cekirge, H.M. and Raza, S.A.R. (2013) An assessment of the potential contribution from waste-to-energy facilities to electricity demand in Saudi Arabia. *Energy Conversion and Management* 75, 402–406. DOI: 10.1016/j.enconman.2013.06.056.

Pourtaheri, M., Rahmani, K. and Ahmadi, H. (2012) Impacts of religious and pilgrimage tourism in rural areas: the case of Iran. *Journal of Geography and Geology* 4(3), 122–129. DOI: 10.5539/jgg.v4n3p122.

Qurashi, J. (2017) Commodification of Islamic religious tourism: from spiritual to touristic experience. *International Journal of Religious Tourism and Pilgrimage* 5(1), 89–104.

Raouf, A. (2016) A green Haj is a must. Available at: https://gulfnews.com/opinion/thinkers/a-green-haj-is-a-must-1.1891527 (accessed 8 November 2019).

Redfern, B. (2010) A Holy City in transition: investment in Mecca is booming as the government seeks to improve facilities and accommodate growing visitor numbers. *MEED: Middle East Economic Digest* 54(40), 44–45.

Rinschede, G. (1990) Religious tourism. *Geographische Rundschau* 42(1), 14–20.

Robinson, N. (1999) *Islam: A Concise Introduction*. Georgetown University Press, Washington DC.

Rountree, K. (2002) Re-inventing Malta's neolithic Temples: contemporary interpretations and agendas. *History and Anthropology* 13(1), 31–51. DOI: 10.1080/02757200290002879.

Sati, V.P. (2015) Pilgrimage tourism in mountain regions: socio-economic and environmental implications in the Garhwal Himalaya. *South Asian Journal for Tourism & Heritage* 8(2), 164–182.

Shackley, M. (2001) *Managing sacred sites: Service provision and visitor's experience*. Thomson Learning, London.

Shinde, K.A. (2007a) Place-making and environmental change in a Hindu pilgrimage site in India. *Geoforum* 43(1), 116–127. DOI: 10.1016/j.geoforum.2011.07.014.

Shinde, K.N. (2007b) Pilgrimage and the environment: challenges in a pilgrimage centre. *Current Issues in Tourism* 10(4), 343–365. DOI: 10.2167/cit259.0.

Shinde, K. (2008) Religious tourism: exploring a new form of sacred journey in North India. In: Cochrane, J. (ed.) *Asian Tourism: Growth and Change*. Elsevier, Oxford, pp. 245–257.

Sigaux, G. (1966) *History of Tourism*. Leisure Arts, London.
United Press International (2010) Hajj attracts 3 million pilgrims. Available at: http://www.upi.com/Top_News/World-News/2010/11/16/Hajj-attractssome-3-million-pilgrims/UPI-21981289926713/ (accessed 8 November 2019).
Vijayanand, S. (2012) Socio-economic impacts in pilgrimage tourism. *International Journal of Multidisciplinary Research* 2(1), 329–343.
Wainwright, O. (2015) City in the sky: World's biggest hotel to open in Mecca. The Guardian. Available at: http://www.theguardian.com/artanddesign/architecture-design-blog/2015/may/22/worlds-biggest-hotel-to-open-in-mecca (accessed 30 December 2015).
Young, G. (1973) *Tourism: Blessing or Blight?* Penguin, Harmondsworth, UK.
Zafar, S. (2015) Waste management in Jeddah. Available at: http://www.ecomena.org/tag/solid-waste-management-in-jeddah/ (accessed 8 November 2019).

11 The Effects of Natural Disasters on Religious Tourism Sites: Earthquakes in the Spiritual Heart of Italy

Michael Di Giovine[1]* and Elisa Ascione[2]
[1]*West Chester University, Pennsylvania, USA;* [2]*Umbra Institute, Italy*

Introduction

In the early hours of 24 August 2016, a 6.2 magnitude earthquake rocked Central Italy. Its epicentre lay deep below small medieval towns on the mountainous border of four regions – Umbria, Lazio, Abruzzi, and Le Marche. The earthquake was so intense that it was felt from Bologna to Naples; one priest from the seaside town of Pesaro, some 260 km north-east of the epicentre, said it lasted so long that he was able to pray the entire *Hail Mary*. Cities renowned for their religious history, such as Norcia in Umbria, were damaged, and the world would soon awake to images of the town of Amatrice in Lazio, known in tourist guidebooks as 'the city of 100 churches', flattened into a pile of dust and rubble. Bearing the brunt of the earthquake, Amatrice lost 236 lives that day; another 51 deaths were reported in Arquata del Tronto and 11 in Accumoli. 'God took them all at once!' one citizen of Amatrice reportedly wailed (Scherer, 2016).

But the earth was not finished with its tectonic upheavals (Figs 11.1a, and 11.1b). Continuous aftershocks further destabilized structures in the region, culminating in two more giant quakes more intense than the first, on 26 and 30 October. They not only destroyed many small towns in the area, but delivered the death knell to the Basilica of St Benedict in Norcia, a significant religious tourism destination for Catholics who visit the birthplace of the founder of the Benedictine Order and European Christian monasticism. Thankfully, no deaths were reported following these two quakes since they occurred during the day and residents were already sensitized to the seismic activity. Rather, images in the international news media focused on fleeing Benedictine nuns and friars amid the rubble of the sacred site. Then in January, when deep snow covered the debris and cracked streets of mountainous settlements with an air of tranquility, one last giant quake hit again in Rigopiano, Abruzzi, sending avalanches that buried unsuspecting villagers alive and trapping tourists for days in ski lodges. It was as if the whole world had come crashing down.

While earthquakes and natural disasters destroy lives, places, heritage sites, and other valuable assets, they are also occasions for the (re)construction and renewal of values. Several religious leaders, including the archbishop in charge of the diocese of Norcia, publicly proclaimed that people had the 'moral duty' to rise again and affirm their identity in the face of adversity as a unified population through

*Corresponding author: michael@michaeldigiovine.com

a

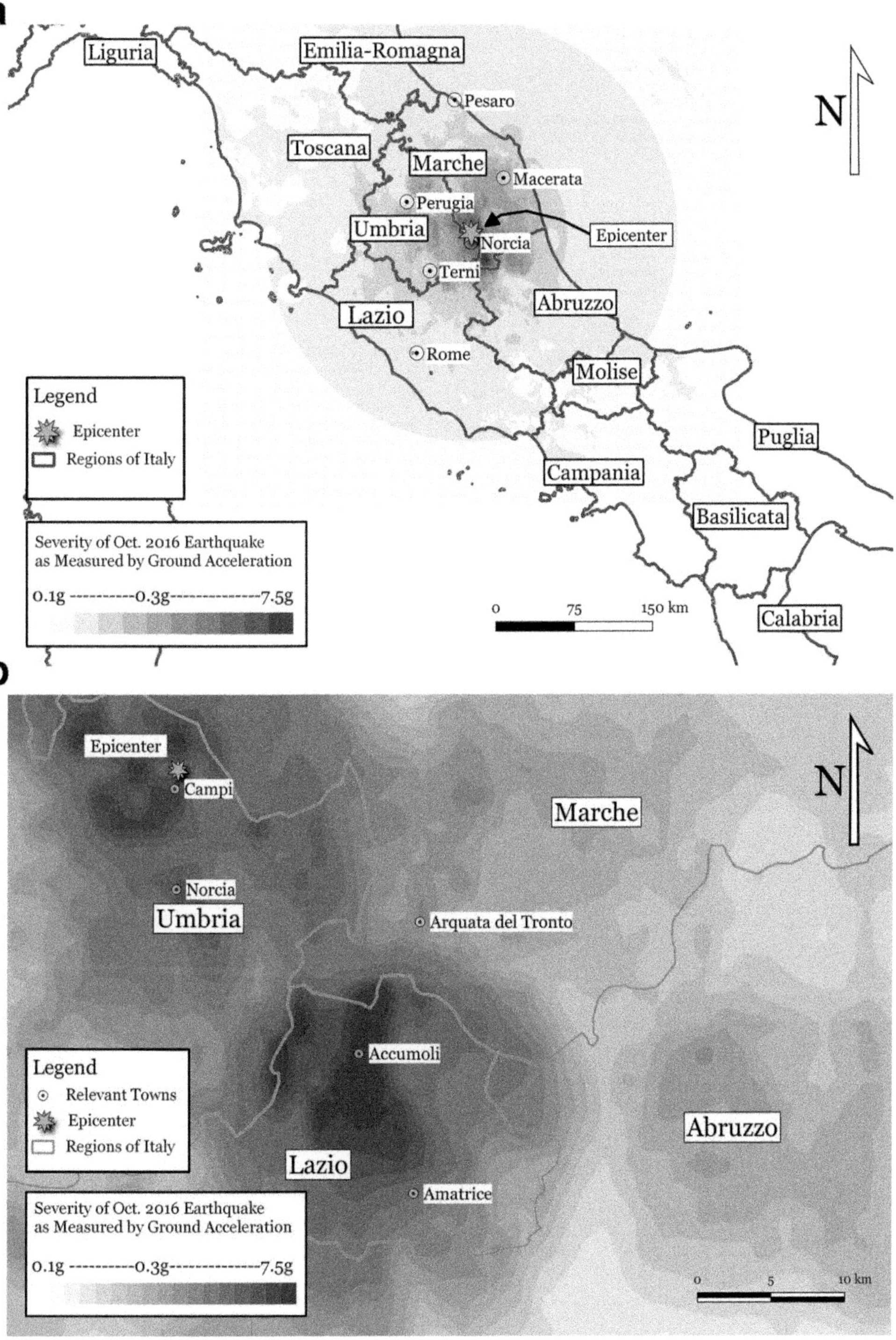

b

Fig. 11.1a and 11.1b. Map of the area affected by the earthquakes (designed by Aaron Gallant).

the religious symbols that connected them (La Nazione, 2016). Locals spontaneously engaged in public acts of solidarity with the earthquake victims, holding public feasts such as cooking and eating Amatrice's eponymous pasta – *spaghetti all'amatriciana* – to raise money and to symbolically show their affection for the community and its heritage (Ascione and Di Giovine, 2016). Tourists and other stakeholders, too, affirmed their presence on the territory as transformative for themselves and for the environment by visiting the affected areas and purchasing goods to support the local economy.

The earthquakes affected many towns that rely on religious tourism, levelling churches in Amatrice, the Basilica of St Benedict within the historic centre of Norcia, and numerous minor sanctuaries of importance to local pilgrims. The earthquakes also cut off the area from the rest of Italy, greatly impacting tourism, its food system, and its population. Yet as the cities attempt to rebuild, modifying the built environment, and the sacred space and the rituals associated with it, tourism – particularly religious tourism – has increasingly factored into economic, moral, and social reconstruction. Sacred sites and their associated pilgrimages are intimately dependent on the environment. On the one hand, as Eliade ([1959] 1987) has famously argued, extraordinary landforms are often perceived as 'axes mundi' or sites of the 'irruption of the sacred' – places in which cosmological perfection bursts through the chaos and disorder of temporal, social life. On the other hand, they offer pilgrims the physical elements of challenge and danger, seduction and sanctification, which characterize the understood sacrifices necessary for salvation (Di Giovine, 2015). However, these theoretical notions largely posit a pilgrimage space as relatively stable, an unchanging container for transformational social activity. But what happens when the surrounding natural environment itself changes, as it did in the months straddling 2016 and 2017?

Based on three years of ethnographic fieldwork including participant observation, site inspections within the forbidden 'red zone' of towns such as Norcia, Amatrice and Campi, and interviews with locals, administrators, tourism industry workers, and clergy, this chapter examines the physical and spiritual impacts of this devastating series of earthquakes in Central Italy, and explores how different stakeholders (e.g. tourist guides, local authorities, shopkeepers, priests, nuns and monks, and tourists) have framed the role of tourism in the area and how touristic practices re-signify the value of places and their associated moral ideals after environmental disasters. In addition, we examine the dialectical effects of a double environmental transformation on sites: how a natural disaster – unplanned and uncontrollable by social actors – impacts the tangible and intangible fabric of pilgrimage sites, and how those subsequent social interventions likewise impact religious tourism and its stakeholders as well as the wider environment itself.

Background: The Sacred Sites of Umbria and Earthquakes

Norcia and Cascia are two towns located in the southern part of Umbria (called the Valnerina, or, 'the valley of the river Nera'). Umbria is a land-locked mountainous region of Italy promoted for decades as a sacred, mystical place, thanks in part to the presence of several important Catholic saints. Norcia was the birthplace of 6th-century St Benedict of Norcia (480–547), the founder of the Benedictine Order. Venerated as a saint by Catholics, Eastern Orthodox, and Anglicans, his famous *Rule of St Benedict* (Benedict, 1998) governing monastic life – with its preaching of moderation, balance, and reasonableness – was adopted by most Christian monastic orders in the Middle Ages, and for this reason he is known as the founder of Western Christian monasticism. St Benedict was also proclaimed the patron saint of Europe in 1967, and in 2008, Pope Benedict XVI lauded the saint for pulling Europe out of the dark ages (Benedict XVI, 2008). A medieval basilica was built over the remains of his childhood house and was later enlarged during the Renaissance era.

Cascia, meanwhile, is the home of St Rita of Cascia (1381–1457), a 15th-century child bride and widow-turned-cloistered Augustinian nun who is revered as a patroness of abused housewives and widows; she pardoned the murderers of her husband and entered the convent after her sons died of dysentery – which was seen as divine intervention preventing them from committing the mortal sin of seeking blood revenge on their father's killers. She is also revered as a patroness of impossible causes; when dying of tuberculosis in January 1457, she asked her friend for a flower, and to her surprise, a single rose bloomed in the midst of the snow. Canonized in 1900, her incorrupt body now sits in a modern basilica in town, drawing thousands of pilgrims annually.

Norcia and Cascia rest in the hillsides and valleys of the Valnerina shrouded by the Sibylline Mountains, so named by the Romans for the sacred Sibyls – divine soothsayers in antiquity who delivered prophesies to pilgrims that would visit them in their dark caves. Towering over them all is the Grand Sasso, or the 'great rock', the highest point in Italy near which L'Aquila, the capital of the neighbouring region of Abruzzi, sits. In 2007, L'Aquila was destroyed by an enormous earthquake, its Baroque edifices – which themselves were the product of reconstruction after an equally cataclysmic series of quakes in the eighteenth century – crashing down and dispersing the population.

Valnerina's landscape is characterized by tiny, entrenched hamlets of medieval origin scattered on top of hills, and abbeys surrounded by woods. It currently has 20,000 residents divided in 14 main municipalities which include a multitude of small villages (Agenzia per la Coesione Territoriale, 2018). From a tourism perspective, the Valnerina is rich in naturalistic attractions, such as the Castelluccio Plain, famous for its picturesque wildflowers in July, and the River Nera with a flourishing rafting industry, woods and foothills for hiking, and a significant production of high quality foods recognized throughout Italy, such as prosciutto di Norcia, black truffles, lentils, roveja (a wild local pea), saffron, and extra-virgin olive oils. The area, as the site of the first Christianization of the region thanks to the settlement of Syrian monks (Agenzia per la Coesione Territoriale, 2018), is also popular for its ancient churches and abbeys, such as Sant'Eutizio, San Pietro a Valle, and Santi Felice e Mauro, and therefore has seemingly always been a land linked to spirituality.

Earthquakes have been very common in Valnerina, which has one of the highest levels of seismic activity in Italy. In the last half-century, Valnerina endured earthquakes in 1979, 1984, 1997, and the last ones in 2016–2017. The 1997 Umbrian earthquake was particularly infamous, as its epicentre was north of the Valnerina under the town of Assisi, one of Italy's most important religious tourism destinations. Assisi is the birthplace of St Francis, the patron saint of Italy and one of the most prayed-to saints in the world (for Italy, see Bobbio, 2006). The Basilica – a UNESCO World Heritage site – is filled with inestimably valued art, such as the fresco cycle, *The Life of St Francis* by the 15th-century painter Giotto, which, art history textbooks proclaim, signalled the birth of Renaissance art (UNESCO, 2000). The 1997 earthquake killed 11 people and collapsed the roof of the Upper Basilica, damaging Giotto's frescoes. During that time, the government acted swiftly, thanks in part to the Basilica's fame and out of the necessity of securing the structure so that it would not fall in on itself, collapsing the Lower Basilica and destroying important Medieval-era artwork by painters such as Cimabue and Simone Martini. Such unusually swift reconstruction by the notoriously bureaucratic Italian government also could have been a result of the timing. The European Union had formally been instated a few years earlier, and Italy was in the midst of negotiating its integration into the Eurozone. As such, it was incumbent that the Italian state demonstrate that it had the economic and intellectual resources to rebuild a site of pan-European importance. In just three years, the Basilica was open again, earthquake-proofed with cutting-edge technology used by NASA (Gayford, 1999), and the art restoration effort was lauded for its authenticity and integrity despite such major damage (ICOMOS, 2000). Religious tourism also spectacularly rebounded. While in the year following the earthquake Assisi saw an 81% decrease in tourist arrivals – from 550,000 to 9000 hotel customers (Caleffi *et al.*, 2000) – and an estimated loss of some $71 million dollars (Mazzocchi and Montini, 2001, p. 1045), by 2017, Assisi boasted over 1.7 million tourists who stayed overnight in hotels – nearly a 120% increase 20 years after the earthquake – and an overall visitation number of 6 million tourists per year (Regione dell'Umbria, 2018), which is an increase of 17% from when the earthquake hit (see ICOMOS, 2000).

Yet the government's intervention was not limited to Assisi alone; it also provided financial assistance to the Valnerina whose towns also suffered significant damage in the 1997 quake. The reconstruction effort that followed helped the Valnerina to transform itself. Thanks to funds allocated by the State, municipalities and home-owners were able to restore buildings that had never been renovated before, bringing them up to date and outfitting them for tourists. For example, 'Dario' – a local resident who, along with his siblings, inherited a 16th-century farmhouse on the outskirts of the village of Campi that they converted into an agritourism business – could completely earthquake-proof his property. He removed the 1960s-era concrete roof and installed a wooden one, remade the thick outer walls with a flexible steel netting inside that would better move with an earthquake, and installed piping into the foundations that absorb seismic shocks.

Although destructive to the Valnerina, paradoxically, the 1997 earthquake helped the region to focus on tourism development thanks to the beautifully restored hamlets and fascinating landscape. The regional government also pushed many municipalities to adopt tourism diversification efforts. While Norcia, Cascia, and other towns were widely known for their religious heritage, they began a dedicated effort to promote agritourism businesses and gastronomic tourism with initiatives such as the festivals celebrating saffron and the Norcian black truffle, the promotion of local lentils and highly prized salami and prosciutto from Norcia (which, revelatory of their fame, are called *norcineria* in Italy), and the development of the Ecomuseum of the Apennines (cf. CEDRAV, 2019). The integration of the area into wider tourism development initiatives is a relatively recent phenomenon. By the late 1990s, many farms in Umbria ceased being productive as the younger generation moved away to the bigger cities in search of university education and work. Yet through funds from the European Union, which were disbursed by the region, many of these educated descendants returned to the area to refurbish their ancestral farms to accommodate agricultural tourism. The 1997 earthquake thus afforded the opportunity for the Valnerina and surrounding earthquake-hit areas to better capitalize on this burgeoning form of tourism through the restoration of religious and historical sites.

Fig. 11.2. A road and house in Amatrice, 2017 (Michael A. Di Giovine).

Rethinking Tourism and Communities After the 2016–2017 Earthquakes

Twenty years after the Assisi earthquake, the situation was far different from that of 1997. Rather than having occurred on one day and concentrated in specific iconic places like Assisi or L'Aquila, the series of quakes in 2016–2017 tore through the region for nearly half a year, completely destroying many small towns and villages that were not well-known pilgrimage routes, but rather of more local import. Amatrice, which sits in the Lazio region not far from Valnerina, was arguably the most renowned, thanks to its election into the prestigious Borghi Più Belli d'Italia consortium (the consortium of the 'most beautiful villages in Italy') in 2015 – a designation that celebrated its picturesque setting and its fresco-filled Cento Chiese (100 churches). Known for its *amatriciana* pasta, Amatrice would celebrate its famous dish in a summertime *sagra*, a food-filled festival that is a significant draw for domestic tourists from the surrounding communities (see Di Giovine, 2014) (Fig. 11.2).

Except for Norcia and Cascia, Amatrice and the other villages hardest hit during the earthquake were not iconic tourism destinations – religious or otherwise – but factored into more informal VFR (Visiting Friends and Relatives) tourism practices which would accommodate seasonal flows of kin from the big cities of Rome, Perugia, and the northern regions of Italy who had left in the 1960s to look for jobs but often maintained their ancestral homes in the village (see Aramini, 2013). *Sagre*, as well as the feast days for the town's patron saint and other religious holidays, were occasions where the dispersed population from across Italy would come back together, making pilgrimages, participating in devotional processions and Masses, and cooking and eating the gastronomic fare of their homeland. These small towns, then, can be seen to engage not in major religious tourism or pilgrimage like Assisi, Norcia, or Cascia, but rather in more localized forms of what can be called 'religious VFR tourism', an integration of VFR tourism with strong religious (and gastronomic) components.

The 2016–2017 earthquakes were more diffuse that earlier ones, too, creating significant destruction across a wide swathe of land and blocking the narrow, mountainous roads that would lead to these little hamlets. The earthquake materially destroyed not only the religious centres that were destinations for such travel but also people's homes. Communities were displaced as the Italian government sought to protect the people first. When interviewed about the reconstruction efforts of the 100 churches of the town, Sergio Pirozzi, the outspoken mayor of Amatrice at the time stated to the Italian news service RAI that 'the aim now is to save as many lives as possible'; the town's officials would think about reconstruction efforts later (Dettmer, 2016). Those who survived the earthquake either returned to their primary residences in the big cities or left to stay with family members, while the Italian government provided temporary housing outside of the earthquake zone for those that did not have this social safety-net. Residents of Amatrice were put up in seaside hotels in neighbouring regions, creating serious ruptures in the social fabric of the town as well as psychological and emotional tensions. When we asked Pirozzi what he would like the government to provide first, he replied, 'Psychologists. We need psychologists more than anything now' (fieldnotes, 3/6/2017).

Indeed, Dario, the owner of one agritourism business just outside of the small town of Campi, was traumatized by the earthquakes. He remembers the first thing he thought as the quake started was, 'It is finished' – he was going to die. He still has vivid memories of that day, the trauma of which continues to haunt him:

> After the earthquake I cannot sleep anymore; I suffer from sleep deprivation. That day I heard all the furniture falling. I sat down waiting for it to end. Then I phoned the guests that had to come to my agritourism business and I told them not to come. I jumped into the car to go away from the agritourism since I was scared; I felt the second shake as I was driving. The car was moving by itself. Then there was a total blackout in Campi. An old man knocked at my car window and cried: 'Get out of the car, the earth is coming down!' I put my feet on the road, and it felt like the road opened up in two halves. So I tried to go back to Campi, the houses were falling down on the street. The morning after people called me crying saying: Campi doesn't exist anymore (fieldnotes, 20/5/2018).

Rebuilding so many little towns and edifices would be incredibly taxing and cost-prohibitive. For example, between the now-destroyed upper city of Campi (referred to by locals as Campi Alto) and the lower city of Campi (Campi Basso) sits the small church of St Anthony Abbot, part of a late 15th-century Benedictine sentry and hospital system that hosted, protected, and helped cure pilgrims. With Renaissance inscriptions and frescoes, the church served as a heritage site and a point of reference for the residents of the two villages. The edifice was severely damaged during the earthquake; its roof, side walls, and beautiful rose window caved in. When it was clear the government would do nothing, residents appealed to the Catholic Church. Apparently, according to a resident, the Church determined it was of too minor import to warrant reconstruction. '*Meglio buttarla giù*,' the Church official apparently said; 'it's better to just knock it down.' Community members persisted and succeeded in having the government stabilize the remaining façade and crumbled walls with steel girders and built a tin roof at the cost of upwards of EUR 100,000 (Fig. 11.3).

In addition, many of the villages like Amatrice and Campi boasted a high rate of absentee owners whose primary residences are elsewhere. Their properties were considered second homes and therefore not eligible for state-sponsored reconstruction funds, which, since Italian government's coffers are not deep, were only earmarked for the emergency reconstruction of primary residences. While many retained their ancestral homes as a retreat during the hot summer months in Rome, others had sold them

Fig. 11.3. Church of St Anthony Abbott, Campi Alto, June 2019 (Michael A. Di Giovine).

to foreigners; Campi itself boasts homeowners from Holland and Germany. Indeed, only two extended families lived in Campi full-time. This has complicated reconstruction efforts, particularly since the government has instituted a '*piano integrale*' for reconstruction – a holistic, integral plan in which all of the town (or none of it) must be reconstructed all at once. On paper, admits one tourism entrepreneur, this makes sense: the urban environment and its surrounding areas is indeed all one system, and thus should be reconstructed together (fieldnotes 23/6/2019). People could not get into their homes or parish churches without first clearing the narrow streets of debris, and even when they could reconstruct their property, it would be vulnerable to surrounding structures that were not reconstructed. Piecemeal reconstruction would pose a public safety hazard, exacerbated by the lack of electricity and sewage treatment that would have to first be provided to the whole town. In practice, this has resulted in a standstill for several reasons.

On the one hand, as noted earlier, there were a number of second home owners who did not qualify for public funds. Requests by these arguably wealthier second home owners to fund the work privately are denied for the same reasons: the primary home owners need to have their funding first, and there may not be enough money to go around. On the other hand, small towns are often sites of deep-rooted family rivalries – a tribalism of sorts, which play out in community negotiations for reconstruction. One Dutch couple who own a home in Campi and wanted to use their private funds to rebuild it were exasperated at a town meeting: they asked the citizens to band together to restore their houses, but there were some who refused to do so just to prevent their rival neighbour from being able to rebuild his. 'I won't rebuild my home because his grandfather drove over my grandfather's chicken coop with his tractor, one said ... Really? When did this happen? 70 years ago? You really won't do it because of that!?' (fieldnotes, 24/6/2019).

Thus, Norcian associations, residents, restaurateurs, and shopkeepers who were especially reliant on religious and alimentary businesses started 'the white bed sheet movement' (*movimeto lenzuoli bianchi*), organized by the Comitato Rinascita Norcia (Committee for the Rebirth of Norcia). Created two years after the earthquake, on 30 October 2018, the Committee noted that the reconstruction work of rebuilding houses, jump-starting economic activities, delivering essential services such as health care and functioning hospitals and schools, and the preservation of public monuments and cultural heritage properties was proceeding very slowly, creating widespread distrust among the population who unfortunately saw the recovery and revitalization of their territory as increasingly unattainable. The social and economic lives of local residents were affected by these delays, which, together with the closure of the road to the Marche region, began to lead to the abandonment of the territory, especially among entrepreneurs and young people, risking severe depopulation of the area (Tutto Oggi, 2018).

Such displacements have taken on a moral ethos, as well. One man pointed out that while the English language has two terms for the place in which one lives – 'house', meaning the structure; and 'home', meaning the place in which one feels safe, connected and centred – Italians only have one term – *casa*, or home; it is both the material structure and the immaterial qualities of personal sacredness that lie within one's heart. This sentiment echoes Eliade ([1959] 1987), who argues that humans recapture and attempt to replicate sacredness not only through the construction of cult buildings such as temples and churches, but quintessentially through their homes; the home is where the sacred is most intimately felt, and where individuals connect with it on a more frequent level – where they escape into their own idea of perfection amidst a difficult and imperfect outside world. 'And so it just kills me to see these older people, knowing that they will never return home again,' a local man sighed (fieldnotes 24/6/2019).

Yet while the earthquake has created very materially instantiated social divisions, it has also paradoxically created forms of solidarity among some residents, reformulating the role of tourism and the sense of community. For example, the residents of Norcia – like those in Amatrice – were given 24 hours to board buses that would take them out of the Valnerina indefinitely. However, a group of more radical locals calling themselves the 'Montanari Testoni' ('the hard-headed mountain people') called on their perceived personalities of being tough and obstinate – qualities given to them from the rough mountain terrain (even though Norcia is in a valley) – and refused to move, creating instead a small community of trailers and shipping containers right outside the gates of the town. Showing us their little café, makeshift piazza, and cinema-cum-library where the few residents could relax, the leader of the group said that their grassroots movement demonstrated how, paradoxically, the earthquake could create an opportunity for people to experiment with more democratic ways of interacting and working together – how associative life can generate opportunities to rise again as a revitalized community. After the earthquake, for example, the leader got to know people from the area that she had never met before; she held meetings with elderly residents who possessed deep knowledge of the territory and whom she felt should have been consulted in the reconstruction phase. She realized that some segments of the population had been silenced and marginalized in the race for funding that mobilized different interest groups who did not always act in the general population's best interests. Indeed, as has been shown in other earthquake contexts in Italy, catastrophes can create an 'organic crisis' that triggers new forms of (re)existence with regenerative potential (Pitzalis, 2015, 2016).

From a tourism perspective, Norcia has suffered greatly. In the words of one resident, 'the economy [is] finished' in the city of Norcia, since people were evacuated, buildings fell down, and the majority of churches crumbled (Fig. 11.4):

> The economy stopped the day of the earthquake. Norcia is finished, I live there. Norcia for 20 years has focused on reconstruction and has transformed itself, starting from Mr Rossi, who owns many hotels and restaurants in town, and has put a lot of effort in the requalification of the city. Even in winter, they used to organize many events to attract tourists, and now after the earthquake, unfortunately for them, nobody goes there anymore. I used to go and have dinner on the main street before, but now I don't go there anymore, even though a couple of restaurants have opened up again. During the day there are some people, but at night it's like a ghost town, there are no more tourists. There's nothing left to see in Norcia. Castelluccio is all broken, too; there is just some street food left (fieldnotes, 20/5/2018).

Many shops had to close and were relocated to containers outside of the town centre, organizing into associations to coordinate their actions and to find unity in the disaster. Locals created the consortium 'We are Norcia' in an effort to combine their shops, restaurants, and agritourism businesses under one logo and to increase online sales and attendance at fairs, events, and even political round-tables, with the common aim of bringing life back to the area. 'Norcia will restart again, thanks to the willpower of its people, thanks to its territory, its beauty and strength. Come and visit us, be a witness to our victory!' they state on their website (www.wearenorcia.com).

Witnessing and Solidarity Movements

Witnessing is a common term in Christianity, and particularly in American Protestantism. Witnessing denotes not a passive watching, but an active engagement with, and performance of, spiritual values (Elisha, 2011; Dwyer, 2017). One 'witnesses' the power of God through good

Fig. 11.4. Piazza San Benedetto, Norcia, 2017. The iconic statue of St Benedict stands in front of the destroyed Basilica of St Benedict (under scaffolding). The scene remains the same today (Michael A. Di Giovine).

works and charity (*caritas*), to use a Catholic term. While these residents, who would identify primarily as Italian Catholic, may not have used this term explicitly with such theological conceptions in mind, the idea that tourists can be 'witnesses' and helpers of the local population was repeated by many of our informants. Dario, who sometimes even encourages his current guests to hike up to the edge of the destroyed town of Campi, said, 'I hope that the tourists that I host here – sleeping, eating, and shopping here – become witnesses of the work that we do in order to be back on our feet. Tourists are crucial for this' (fieldnotes, 20/10/2018). After we carried out an inspection of the red zone on another occasion, he said, 'It's sad, right? But it also can bestow hope. It can give witness to the human dimension, to what has happened, and what still needs to be done' (fieldnotes, 23/6/2019).

When evaluating the ways in which tourists and others can witness, or at least help out the population, solidarity is another term that many residents utilized. Locals focused on two types of solidarity. The first and initial form is what can be considered 'remote solidarity' – that is, supporting local enterprises economically from afar. The 'We are Norcia' campaign, for example, focused much of its efforts on promoting the gastronomic products for which the city is famed. During the Christmas season of 2016, after Norcia's Basilica collapsed and its prosciutto producers' storage facilities, animal pens, and farmhouses were destroyed, local entrepreneurs successfully conducted an online campaign to purchase *norcineria* as Christmas presents. This initiative veritably saved the Norcian meat industry: the Christmas gift baskets sold out. The demand had been so great that the following summer, norcineria producers at markets in Perugia had notably lower reserves to sell.

This inspired the 'United for Campi' campaign, which similarly includes a box with foods from local producers, such as cured meats, cheese, marmalades, and barley that people can buy in order to support the local economy. They sold more than 500 boxes in a year, according to Dario, who was also one of the prime promoters of the initiative. Part of the money was donated to the Pro Loco di Campi, a local association that works to promote the area; most towns have Pro Locos, and they are typically occupied with tourism promotion. There, the president of the Pro Loco has built a 'solidarity area' in a public space in which people slept as they waited to be officially relocated, where they would hold meetings to discuss future plans. They hosted 70 people for six months, thanks to the Protezione Civile that brought beds to the area. Calling it a 'great example of community,' Dario said:

> The help of people was amazing: we discovered this thanks to the earthquake and it warmed our heart. Private help arrived where public help didn't arrive. I understand that the 2016 earthquake was unexpected, it's not easy to operate and to reach these areas, when mountains collapse. But now it's 2018, two years have passed. They still need to do so many things: the stones of the church of Campi that fell on the street are still there, untouched. I personally moved them with my neighbours, otherwise the road would still be blocked (fieldnotes, 21/8/2018).

There have emerged many such examples of remote solidarity towards the affected population, where people purchase local goods or donate funds from afar to help sustain the vulnerable populations of these towns (see Ascione and Di Giovine, 2016). However, it was also crucial that people show their support for the local residents through what can be called 'embodied solidarity' – the physical act of being there, helping to bring back tourism in the area by 'witnessing,' and by continuing to visit the affected areas whenever possible. Many entrepreneurs told us that each time an earthquake hit, last-minute cancellations spiked; and those who did come spent relatively less time than they used to:

> Now I have 45% less of tourists. We were 24 agritourism businesses before the earthquake, now only six of us opened again. We had eight hotels in Norcia, and only one is open now. The typologies of tourists have changed: we have lost families, 40–50 years old, with children, we don't have them anymore. That was 80% of my tourism: they came for historical tours, to see the hamlets, then 20% came for environmental tourism, mountain bikes, hikes, etc.. Another effect of the earthquake has been that people now stay for only two days, not anymore for four. Now the territory must reinvent itself with nature and landscape tourism because there's nothing else to do now, the majority of the churches and art was destroyed (Luigi, business owner, fieldnotes, 24/6/19).

Yet it 'warms the heart' when he receives calls from repeat clients asking, 'How are you doing? I'll come and stay just to be there for you' (fieldnotes 24/6/19). As one tour operator of the area told us, the future recovery of the Valnerina is inextricably linked to the development of a 'conscious tourism' that takes advantage of, but also respects, nature, landscapes, and religious sites.

However, tourism by itself cannot help the revitalization of the area, and there is a general consensus that institutions must provide solutions for local residents to keep them from permanently leaving the Valnerina. Our informants claimed that they need more hospitals, schools, and services in order to keep residents in the territory. Campi, for example, had 120 residents during the year, but swelled to 500 during the summer when the second home-owners returned. Tourists visited because they loved the landscape, but also because they enjoyed the social life of a small village. As a local resident told us:

> We live in a place where we leave our car unlocked; these are the small details that make the quality of life of this place. When tourists come here, they love this way of living. We must preserve this, because people were already leaving mountain areas even before the earthquake, and now so many want to go away from here (fieldnotes, 21/5/2018).

For tourism initiatives to be successful, it is thus crucial that local people have incentives to stay, even those who do not directly work in the tourism sector. Indeed, many local youths have been moving to other places to look for jobs after the earthquake. This type of 'flight from the land' (cf. De Kadt, 1979) is troubling: tourism itself is precarious and fickle, and most of all it is interconnected and highly dependent on the social life of the local population. In the re-imagining of the role of tourism in the recovery process, religious sites like Cascia and St Rita's birthplace of Roccaporena are very important, but so is the integration of non-religious forms of tourism focusing on cuisine and gastronomic practices. This diversification, furthermore, benefits not only the tourist, but, if holistically integrated, serves as a valuable method for the exchange of best practices and the creation of unity among local entrepreneurs. As a hotel manager in Cascia stated:

> If we want to make this territory alive again we must link everything to tourism – saffron, agricultural activities, cheese making, and religious tourism must be a part of this. We must unite among hotel managers, for example, unifying the prices and sharing ideas (fieldnotes, 23/10/2018).

Religious tourism has not only shown resilience in the face of adversity but also has undergone a transformation by becoming more integrated with cultural, natural, and food tourism. In the face of adversity and in moment of crisis, economic exchange of goods and services among hosts and guests can become imbued with symbolic meanings and be used as means to bridge the distance between people that are suffering and those who are not. As shown in different crisis contexts, such as in Greece after the imposition of European Union austerity measures, solidarity networks are simultaneously an offshoot of the crisis and a means to contain some of its most dreadful results (Rakopoulos, 2016, pp. 4–5). The solidarity economy is not just a mechanistic reaction to the hardships people encounter due to crisis, but also a way for people to rethink their roles as part of a larger 'humanity'. Part of this larger sense of self is achieved through the circulation of bodies (tourists), food, and money.

Reimagining Religious Tourism in Broader Contexts

The statistics from the region's Office of Tourism show that visitation has significantly declined after the earthquake: Norcia hosted 84,136 people in hotels during the first eight months of 2016, and only 10,228 during the first eight months of 2018. The total number of tourists to Norcia (including day trippers) in August 2016 was 116,195, and declined nearly 79% to 24,703 in August 2018 (Regione dell'Umbria, 2018). However, Cascia, the other major religious tourism site in the Valnerina, only experienced an 11% drop, from 131,859 in 2016 to 117,211 in 2018. Religious tourism to Cascia was not significantly altered by the earthquake, since the damage was mostly to the new part of the town, not the old part where the sanctuary is located. Furthermore, the local Office of Tourism decided not to overemphasize the earthquake in their communication strategy to avoid scaring people. Indeed, the role of the media has also impacted this: with its focus on the destruction of Amatrice and Norcia, Cascia has largely been left off the touristic imaginaries (Salazar and Graburn, 2014) associated with earthquakes and danger in the region.

Diversification is key for Cascia's tourism industry, since it is visited by different demographics of travellers who, despite sharing some interest in St Rita, have diverse needs and abilities. According to tour operators in Cascia, 80% of tourists are Italian and come mainly from Southern Italy, where popular devotional practices are quite common (see Carroll, 1992). They include a mix of elderly visitors, young people, couples, organized groups, and individual visitors, with a roughly a 70% return rate of visitors to the region every year. During the interviews, operators described several different categories of pilgrims: one-third are FITs ('free and independent travellers') who do not travel in large groups. They include retired people as well as young couples who came to Cascia with their grandparents when they were children, and then, once married, visit the saint again. The other two-thirds of pilgrims, as one hotel director told us, are organized groups. Within groups, more women than men go to the sanctuary 'because men live their spirituality differently, women are more open to devotion', according to one tour operator (fieldnotes, 23/10/2018). There are also different patterns of visitation. Groups typically visit Assisi, Cascia, and the nearby sanctuary of Collevalenza, which houses the tomb of a mystic nun who is currently being considered for sainthood. But unlike group travellers, FITs have a diversity of interests and tend to stay longer, with visits to Norcia and its food producers.

Yet Cascia, as a site of an internationally important saint, also hosts pilgrims from all over the world, and their different qualities and needs are described in this way by a tour operator:

> The people from the South of Italy are exalted just by looking at the church; the Central and

> Northern Italians are much quieter. Then we have African groups; they are very folkloric, they put on dresses with St Rita prints, they sing and they dance – it's very beautiful. Then Belgium, France, Ireland, they live their spirituality more intimately. Then Argentinians, Brazilians, Lebanese – the statue at the entrance of the village has been donated by a donor from Lebanon. Polish people are very quiet; they're very silent. We have the all world here in Cascia! (fieldnotes, 25/11/2018).

'Purely religious' pilgrims are considered to be those devoted to St Rita. Traditionally they would only visit Cascia and Roccaporena, and are generally considered to come from lower economic backgrounds than the hikers. Although mainly interested in the visit to the holy sites linked to Rita, their expectations and desires have changed throughout the years. Hotels had to upgrade their offerings to satisfy them and attract them, such as proposing à la carte menus rather than fixed priced meals or new activities:

> People now seek comfort anyway; in the past pilgrims maybe stayed in six people in a room, and when they came to Cascia for the Santa they were happy even just to be in a hotel. Now people have everything in their homes, so when they travel they also want [the same] comforts. Religious tourism was essentially a poor people tourism; they didn't want modern environments, so the hotels did not really rush to modernize, so we fell asleep a little bit as tourist providers. Now we're paying the price for that, and we must rush to make new investments, because people have changed (hotel director, interview, 25/11/2018).

Although in the past different kinds of tourism in the area have never really been integrated, more recently operators have been trying to use religious tourism to propose a variety of experiences, and the earthquake itself had the unexpected effect of making people more aware of the territory around the sanctuary:

> Food, wine and territory: we try to make all of these aspects overlap. Peoples' expectations have changed; only 50 years ago people would come, go to Mass, and go back [to the hotel] to sleep. Now they also want to enjoy good food, the cultural events, they visit the villages. Our vision is to try to let people sleep longer, otherwise they would only stay for one night, mainly at weekends. People come for Santa Rita, they go to Roccaporena, but they don't stay longer. Instead now we try to propose also Norcia. The earthquake was the end of the world for Valnerina, but talking so much about it has also put the curiosity into people who have realized that it's full of beautiful places around, and now they tend to visit them more than before, they have actually rediscovered the territory. In the disaster, we have seen a positive effect (hotel director 2, interview, 25/11/2018).

Religious tourism has therefore become the main driver of the local economy. As one person in Cascia told us, 'without St Rita, there wouldn't be anything here' (fieldnotes 25/11/2018), since the cafés, shops, B&Bs, hotels, as well as the 200 people employed by the monastery, depend on pilgrims. Furthermore, the nuns' successful fundraising activities have allowed them to build hospitals and schools in Cascia. Of course, just after the earthquake, people left the Valnerina, and it took almost a year to have an important flux of people again.

> As soon as the earthquake happened nobody came, everybody was scared. When we had the first earthquake in August our hotel was full, and everybody just went away on that night. I have to say we had a lot of solidarity from all over the world, boxes, money, people donated a lot to the monastery. We didn't work at all for a while; the historic groups cancelled – I understand it. There are groups that come every year, some groups come every year for 20 years, [and] they always book one year in advance (hotel director 2, interview, 25/11/2018).

Another important demographic of visitors to the sanctuary are hiking groups that are interested in the walks of St Francis and St Benedict, and who visit the saints in passing. One local tour operator, however, suggested that hikers must be distinguished from the 'real' religious pilgrims who travel specifically for devotional purposes. Furthermore, the naturalistic/religious walks are emblematic products that 'look good on brochures and websites', another tour operator told us, but which unfortunately do not have a strong impact on the economy. 'People want to artificially recreate the experience of pilgrimage by spending as little as possible,' often without showing 'real' devotion towards the saints, he lamented. Although it seems like a more frugal way to travel, only people who can afford to stay away from home for many days can do these long hikes, unlike most tourists who only stay for one or two days in Cascia, solely for St Rita.

The St Benedict trail is 300 km long and connects Norcia to Subiaco. It was created to popularize St Benedict's main sites, allowing people to discover more isolated, lesser-known hamlets that are only marginally touched by tourism. According to its creator, who is a member of the Scout movement (similar to the US Boy and Girl Scouts), any person who walks in search of something, even if not for a religious quest, can be called a pilgrim. Yet people do not walk the St Benedict trail solely out of devotion to the saint, but also for a variety of reasons, such as an enjoyment of hiking, a love of nature, or a sense of accomplishment. After the earthquake, following the trail acquired an even deeper meaning for hikers: to show embodied solidarity to the people living in the affected areas. In March 2017 and March 2018, more than 100 hikers participated in the '*Cammini della Solidarietà*' (Solidarity Walks) to help the local economy, bringing a sign of closeness to the affected population. Thus, on the one hand old tourist routes have been resignified after the earthquake, and on the other hand a new kind of post-disaster tourism has emerged, as evidenced by the recent publication of the guidebook *Il Cammino delle Terre Mutate: Trekking di Solidarietà* (*Walks in Transformed Environments: Solidarity Hikes*; Sgarella, 2019), which includes a map for a 14-day hike from Fabriano to L'Aquila, passing through Valnerina. Locals do not seem to perceive it as a form of 'dark tourism' (Sharpley and Stone, 2009; Olsen and Korstanje, 2020) – tourism to sites of death and destruction, which is often criticized as being voyeuristic and superficial in its minimization of real suffering. Instead, it seems to be perceived as a way of getting closer to 'the wounds of people', 'tiptoeing' in their sites with a low-impact, softer tourism (tour operator, fieldnotes, 25/11/2018). The post-disaster initiatives which developed in relation to tourism highlight how post-quake tourism narratives can transition from narratives of destruction and loss to narratives of renewal and hope (Tucker *et al.*, 2016).

Nevertheless, many locals have also required spiritual, as well as psychological, attention. The Father Rector of the Basilica of St Rita explained how the earthquake has provided an important moment of reflection for the entire population. People had to ponder the fragility of earthly possessions and of human lives, having to develop some form of reciprocal help and collaboration, often in successful ways. His role as a spiritual guide was very important for the local population during that time. He made sure that locals and tourists could continue accessing the places linked to St Rita, such as being able to pray in the square in front of the Basilica before its post-quake restoration:

> People needed to feel that there was still a community here; it was important that people didn't run away. So I talked to them about patience, about hope, about going back to life again, and we got so much help from all over Italy: from cities, associations, the scouts, the civil protection, from Sardinia and Northern Italy. The civil protection built a camp in Piazza Leone XIII to host people who didn't have the means to sleep somewhere else, or who didn't want to go away to Perugia. My old friends, priests in other parishes, gave us all of their donations. There were many initiatives: the Misericordia [civil protection group], which is voluntary, was very active here in Cascia, cooking 3000 meals per day for everybody. Slowly, we started life again, people came to visit St Rita again, thanks also to the reopening of our church. Then shops opened again, and so on. We learned to collaborate and share – not always; we also encountered people that were very close minded – that's the message that the earthquake gave us: we must transform this into a moment of consciousness, thinking about the fragility of our possessions, but also encouraging [us] to build a better society and community, more united, more together (fieldnotes, 29/11/2018).

Indeed, solidarity has clearly taken many different forms in response to the earthquakes. One interesting example of remote solidarity has been the donation of lambs and sheep from Sardinian shepherds to the shepherds of Valnerina near Cascia. This traditional exchange practice is called '*sa paradura*' in Sardinian dialect, and it consists of lending lambs and sheep to shepherds in difficulty, that then must be returned whenever possible. After the earthquake, 1000 sheep were brought from Sardinia, Nuoro, and then returned to Sardinia a year later under the spiritual guidance of the Father Rector of the Basilica.

Cascia was somehow privileged because the main Basilica suffered limited damage; 'Thank God Saint Rita is OK!' an employee of the Tourist Office told us. The Basilica also received a substantial private donation from the MSC cruise company that allowed for a very fast restoration. But even though religious tourism in Cascia has recovered quickly, members of the clergy and tour operators nevertheless insist on the importance of trying to embed religious tourism in the surrounding natural environment. Tourism needs to be integrative and not be based on solely one aspect, but rather to be diversified. However, this also works both ways: by embedding an individual site in the broader environment and in different industries/areas, tourism risks being 'brought down' with them if something happens.

> Cascia is privileged because it has St Rita, this beacon of sanctity, Rita's presence gave to Cascia an easy way to start again. I think the local authorities should create a 'community of Valnerina', a real association that can offer a variety of things to tourists, and that can help to preserve the environment. The landscape here and the panorama are amazing, we should push towards an environmental tourism to help young people to stay here and not migrate. The greatest fear is the abandonment of the territory; we can't just be happy with religious tourism, people here would be in great difficulty. We should also be attentive to other beauties of the territory, taking care of what the Lord has given to the territory. Let's not abandon it, there are old abbeys, the routes, the walks. The contemplation of this landscape must become part of the religious experience (Rector of the Basilica, interview, 25/11/2018).

Although tourism in Norcia has strongly declined (unlike in Cascia), communities have mobilized around the reimagination of their most sacred sites. The crumbled Basilica of St Benedict has become a symbol of the identity of the town, with tourists taking pictures and pausing around its ruins, and it has been at the centre of public discussions over its reconstruction. The municipality, in fact, publicized a call for proposals from international architects for ideas on what the new Basilica should look like. A few architects have stressed the impossibility of recreating an imagined 'past authenticity', since the shape and structure of the Basilica has been changed many times from its original medieval nucleus, and the latest church was already significantly different from the original. But when news circulated that the remains of the old Basilica might be encapsulated in a larger modern structure of glass and steel, and that a new modern structure would be built outside of the city centre, residents of Norcia rose up. They created an association and made their voices heard during public meetings. One activist, for example, expressed fear that a modern Basilica designed by an 'archi-star' (i.e. an internationally renowned architect) could never be recognized by local people as their own: they did not want Norcia to become 'invaded by mass tourism' like San Giovanni Rotondo in the Puglia region, with its highly criticized, ultra-modern sanctuary for St Padre Pio designed by world-famous architect Renzo Piano (see Di Giovine, 2012). Instead, in their public plea they affirmed their desire for a project that would not create another wound to the city of Norcia, which continues to mourn the old church, by changing the location and style of their most iconic building.

> For us, the Basilica is not just an artifact, an aggregate of stones, styles and works of art. Seeing the Basilica from the Piazza meant looking at a form that was familiar to us, a building on which deep energies converged. Thanks to the Basilica, our pride as citizens of this city, and as religious believers, became tangible (La Repubblica, 2018).

As we have argued here, there is an interconnectedness between holy sites, spirituality, the territory, tourism, and people. Religious tourism has been transformed into a more integrative type of tourism, highlighting the richness of the land and the diversity of offerings – from local food to different types of accommodation – leading to the diversification of tourist experiences. Tourism is perceived as a unifying factor that can help the local economy in the recovery process. Religious tourism and its associated sites are thus of fundamental importance for the local economy, but are also crucial identarian landmarks for local people who have experienced the destruction of their sites as personal wounds and who now want to be included in the reimagination of their city.

Conclusion: Religious Tourism, Earthquakes, and the Need for Social Context

Tourism is a notoriously fickle enterprise, subject to changing economic and political climates, environmental conditions – even as banal as rain or (lack of) snow – and perceived danger, be it human-made or naturally occurring, as the Umbrian earthquakes were. Religious tourism – an integration of touristic processes centred around, or highlighting, the visitation of religious sites – seems to be no different. Indeed, many scholars have attempted to differentiate pilgrimage from tourism (Badone and Roseman, 2004; Di Giovine, 2013a; Mesaritou and Coleman, 2019), with several pointing to the necessary danger and sacrifices that are supposedly built into pilgrimage. Pilgrims, we are told, see themselves as 'anti-touristic' (Di Giovine, 2020) in that they view their activities as existing 'above and beyond' material needs and desires (Di Giovine, 2013b), including, on some occasions, contexts of extreme danger. For example, Arbae'en, the Shia Muslim pilgrimage to their primary shrine in Karbala, Iraq, is one of the largest pilgrimages in the world, drawing upwards of 6 million travellers on foot from Iran and Iraq even during the brutal occupation of the area by the Sunni Islamist group ISIS that threatened their lives (BBC, 2014). Yet as Turner and Turner (1978, p. 20) famously wrote, 'a pilgrim is half a tourist if a tourist is half a pilgrim'; it is indeed often difficult for many religiously inspired travellers to suspend their material fears, desires and needs – as well as their threshold for danger and discomfort – in light of such uncontrollable natural disasters. We have preferred, therefore, to analyse the transformations in visitation practices from the standpoint of religious tourism, which in this case discursively captures better the materiality of these processes.

Furthermore, 'tourists' and 'tourism' were the terms most used by our industry informants. As an emic form of research 'from the bottom up', anthropology places heightened value on the 'native categories' employed by research subjects over those terms and categories imposed by the researcher from the top. Locals and industry professionals seem to utilize the term, too, to emphasize the diversification and multivalence of visitation to the region, marked as it is with an integration of various forms of experiences that visitors often enjoy on the same trip (cf. Terzidou *et al.*, 2018): they may come for St Rita or for St Benedict or a local saint; but they integrate local culinary experiences, trips to natural areas, and perhaps even visits to friends and family as well. These travellers are also clearly subject to serious barriers to visitation, both psychological and material – barriers that are clearly not 'above and beyond' this world. Fear, generated in part from media reports and people's own experiences with earthquakes, result in decreasing visitation in the area as much as do the crumbled buildings and blocked roadways. By considering this religious *tourism*, we are better able to capture the holistic experience of travel to the area as well as the common impacts on the social and economic life of the affected locals which are felt indiscriminate of the typology of (non-)visitor.

This chapter has focused not only on tourism statistics and the direct impacts of the earthquakes on hospitality services or the religious sites themselves, but also on the broader social fabric of these villagers. Locals, who may not have daily interactions with tourists, nevertheless suffered economic, social, and cultural impacts from the earthquakes in the same way as their neighbours who work in the hospitality industry. Their foodways, their sense of home and identity, their physical, psychological, and emotional well-being are equally affected. While this may not always be evident in times of peace and tranquility, a cataclysmic event such as the earthquakes of 2016–2017 reveals how integrated daily workaday life is with that of tourism for all stakeholders. Indeed, as Nogués-Pedregal (2019) points out, one of the things that the anthropological study of tourism has shown us is that to understand the complex nature of touristic processes one must first understand that they are embedded in the broader social life of the actors (see also Leite and Graburn, 2009). This chapter has examined both the effects of earthquakes on religious tourism and religious tourism sites, but also on the actors and associated practices in which they are deeply embedded, so as to emphasize the holistic nature of tourism as well as the synergy between place, people, and imaginaries, particularly in the context of religious travel (cf. Coleman and Elsner, 2003; Eade and Sallnow, [1991] 2000).

Earthquakes, as we have argued, have a destabilizing effect on communities, but are also events that characterize change and mobilize different kinds of work, both material and symbolic. Earthquakes create the need to witness – to provide testimony to, and act on – the power of God and nature as well as the abilities or inabilities of humans to control their environment. They foster diverse types of solidarity – symbolic ways of fostering unity among a community – from 'remote solidarity' that symbolically demonstrates one's connection and support for the affected people to 'embodied solidarity' that leads people to physically connect with those affected by the disaster. They may induce lasting trauma and disaffection with the land, or they may provide heightened senses and perceived connectivity with the environment and its uncontrollable forces. Natural disasters may bring out the worst of sensations – a feeling of abandonment by God, by the government, by tourists, by kin – but can also bring about hope and inspiration, a source to transmit spiritual teachings to a community in need. The perception of sacredness – or cosmological perfection amidst the chaos of human life – emerges equally from the peace and tranquility espoused by the natural and built environment as it does from the upheavals and terror produced by the same during natural disasters. Just as tourism itself is a complex human phenomenon, natural disasters – while largely uncontrollable by humankind – are equally complex and integrative, understood and acted upon in complex and often conflicting human ways. In the end, then, the case of the earthquakes in Central Italy shows us both the humanity in natural disasters and also emphasizes the humanity in tourism.

References

Agenzia per la Coesione Territoriale (2018) Bozza di strategia. 14 comunità una sola idea comune. Ricostruire la strategia di area. Available at: http://old2018.agenziacoesione.gov.it/opencms/export/sites/dps/it/documentazione/Aree_interne/STRATEGIE_DI_AREA/Bozza_della_strategia/Bozza_di_strategia-Valnerina.pdf (accessed 27 June 2019).

Aramini, A. (2013) *La Valnerina Com'era. I Paesi e La Gente, Un Viaggio Tra i.* Frammenti Della Storia. Il Formichiere, Italy.

Ascione, E. and Di Giovine, M.A. (2016) The Amatriciana per Amatrice campaign: reflections on food, solidarity and the earthquakes in central Italy. FoodAnthro.com. Society for the anthropology of food and nutrition. Available at: https://foodanthro.com/2016/10/28/the-amatriciana-per-amatrice-campaign-reflections-on-food-solidarity-and-the-earthquake-in-central-italy/ (accessed 27 June 2019).

Badone, E. and Roseman, S. (2004) *Intersecting Journeys: The Anthropology of Pilgrimage and Tourism.* University of Illinois, Champaign, Illinois.

BBC (2014) Shia pilgrims flock to karbala for Arbaeen climax. *BBC News*, December 14. Available at: https://www.bbc.com/news/world-middle-east-30462820 (accessed 5 March 2020).

Benedict XVI (2008) General Audience: St. Benedict of Norcia. Available at: http://w2.vatican.va/content/benedict-xvi/en/audiences/2008/documents/hf_ben-xvi_aud_20080409.html (accessed on June 27, 2019).

Benedict (1998) *The Rule of St. Benedict*, Edited by Timothy Fry. Vintage Books, New York.

Bobbio, A. (2006) I più amati dagli italiani. *Famiglia Cristiana* 5(45), 66–69.

Caleffi, M., Meroni, F., Montini, A., Zoboli, R. and Zonno, G. (2000) Variation of tourist flows in the Umbria Region after the Septemer 26, 1997, Colfiorito (Central Italy) Earthquakes. *12WCEE*. Available at: https://www.iitk.ac.in/nicee/wcee/article/2518.pdf (accessed 26 June 2019).

Carroll, M. (1992) *Madonnas that Maim: Popular Catholicism in Italy since the Fifteenth Century*. Johns Hopkins Press, Baltimore.

CEDRAV (2019) Ecomuseo. Centro per la documentazione e la ricerca antropologica in Valnerina e nella dorsale appenninica Umbra website. Available at: https://www.cedrav.net/ecomuseo/net/ (accessed 29 July 2019).

Coleman, S. and Elsner, J. (eds) (2003) *Pilgrim Voices: Narrative and Authorship in Christian Pilgrimage*. Berghahn Books, Oxford, UK.

De Kadt, E. (1979) *Tourism: Passport to Development?* Oxford University Press, Oxford, UK.

Dettmer, J. (2016) Central Italy earthquake death toll rises, search for victims continues. *Voice of America News*, August 24. Available at: https://www.voanews.com/europe/central-italy-earthquake-death-toll-rises-search-victims-continues (accessed 27 June 2019).

Di Giovine, M.A. (2012) A tale of two cities: Padre Pio and the reimagining of Pietrelcina and San Giovanni Rotondo. *Textus: English studies in Italy* 1(1), 155–167.

Di Giovine, M.A. (2013a) Apologia pro turismo: breaking inter- and intra-disciplinary boundaries in the study of tourism and pilgrimage. *The Journal of Tourism Challenges and Trends* 6(2), 63–94.

Di Giovine, M.A. (2013b) Sacred journeys as spaces for peace in Christianity. In: Pazos, A. (ed.) *Pilgrims and Pilgrimages as Peacemakers in Christianity, Judaism, and Islam*. Ashgate, Surrey, UK, pp. 1–38.

Di Giovine, M.A. (2014) The everyday as extraordinary: Revitalization, religion, and the elevation of Cucina Casareccia to heritage cuisine in Pietrelcina, Italy. In: Brulotte, R. and Di Giovine, M.A. (eds) *Edible Identities: Food as Cultural Heritage*. Ashgate, Surrey, UK, pp. 77–92.

Di Giovine, M.A. (2015) The seductions of suffering: stigmata, salvation and pilgrimage to the shrine of St. Padre Pio in San Giovanni Rotondo. In: Di Giovine, M.A. and Picard, D. (eds) *The Seductions of Pilgrimage: Sacred Journeys Afar and Astray in the Western Religious Tradition*. Ashgate, Surrey, UK, pp. 187–210.

Di Giovine, M.A. (2020) Between tourism and anti-tourism: ethics and the study abroad experience. In: Bodinger de Uriate, J. and Di Giovine, M.A. (eds) *Study Abroad: Student Travel and the Quest for an Anti-Tourism Experience*. Lexington Books, Lanham, MD.

Dwyer, J. (2017) What it means to witness: honoring our United Methodist vow. The people of the United Methodist church. Available at: http://www.umc.org/what-we-believe/what-it-means-to-witness-honoring-our-united-methodist-vow (accessed 6 January 2019).

Eade, J. and Sallnow, M. ([1991] 2000) *Contesting the Sacred: The Anthropology of Pilgrimage*. University of Illinois Press, Urbana, Illinois.

Eliade, M. ([1959] 1987) *The Sacred and the Profane: The Nature of Religion*. Harcort, Orlando, Florida.

Elisha, O. (2011) *Moral Ambition: Mobilization and Social Outreach in Evangelical Megachurches*. University of California Press, Berkeley, California.

Gayford, M. (1999) The miraculous rebirth of Assisi. *The Telegraph*. 27 November. Available at: https://www.telegraph.co.uk/culture/4719139/The-miraculous-rebirth-of-Assisi.html (accessed 27 June 2019).

ICOMOS (2000) *Assisi, No. 990. Advisory Body Evaluation, June 30, 1999*. ICOMOS, Paris.

La Repubblica (2018) 'Ricostruite la Basilica di Norcia dov'era e com'era': l'appello di un comitato di cittadini. *La Repubblica*, September 8. Available at: www.repubblica.it/cronaca/2018/09/08/news/norcia-205931768/ (accessed 29 July 2019).

La Nazione (2016) Terremoto, il vescovo di Norcia: 'Rinascano i tesori religiosi'. *La Nazione*, October 29. Available at: https://www.lanazione.it/umbria/cronaca/terremoto-boccardo-tesori-1.2635612 (accessed 26 July 2019).

Leite, N. and Graburn, N. (2009) Anthropological interventions in tourism studies. In: Robinson, M. and Jamal, T. (eds) *Sage Handbook of Tourism Studies*. Sage, London, pp. 35–64.

Mazzocchi, M. and Montini, A. (2001) Earthquake effects on tourism in central Italy. *Annals of Tourism Research* 28(4), 1031–1046. DOI: 10.1016/S0160-7383(01)00008-1.

Mesaritou, E. and Coleman, S. (2019) Distinction and devotion. *Historisch Tijdschrift Groniek* 215, 179–192.

Nogués-Pedregal, A.-M. (2019) Anthropological contributions to tourism studies. *Annals of Tourism Research* 75, 227–237. DOI: 10.1016/j.annals.2019.02.005.

Olsen, D.H. and Korstanje, M.E. (2020) *Dark Tourism and Pilgrimage*. CAB International, Wallingford, UK.

Pitzalis, S. (2015) La catastrofe come crisi organica ri-generativa. Per un'analisi gramsciana del sisma di maggio 2012. *Dada Rivista di Antropologia post-globale* 2, 103–118.

Pitzalis, S. (2016) *Politiche Del Disastro. Poteri e Contropoteri Nel Terremoto Emiliano*. Ombre Corte, Verona.

Rakopoulos, T. (2016) Solidarity: the egalitarian tensions of a bridge-concept. *Social Anthropology* 24(2), 142–151. DOI: 10.1111/1469-8676.12298.

Regione dell'Umbria (2018) Flussi turistici in Umbria nel 2018. Available at: http://www.regione.umbria.it/statistiche-turismo-2018 (accessed 27 July 2019).

Salazar, N. and Graburn, N. (2014) *Tourism Imaginaries: Anthropological Approaches*. Berghahn, Oxford, UK.

Scherer, S. (2016) Italy Quake: 'God took them all at once'. Reuters news service, August 24. Available at: https://www.iol.co.za/news/world/italy-quake-god-took-them-all-at-once-2060546 (accessed 27 June 2019).

Sgarella, E. (2019) *Il Cammino nelle Terre Mutate*. Terre di Mezzo Editore, Milan.

Sharpley, R. and Stone, P. (2009) *The Darker Side of Travel: The Theory and Practice of Dark Tourism*. Channel View, Bristol, UK.

Terzidou, M., Scarles, C. and Saunders, M.N.K. (2018) The complexities of religious tourism motivations: sacred places, vows and visions. *Annals of Tourism Research* 70, 54–65. DOI: 10.1016/j.annals.2018.02.011.

Tucker, H., Shelton, E.J. and Bae, H. (2016) Post-disaster tourism: towards a tourism of transition. *Tourist Studies* 17(3), 306–327. DOI: 10.1177/1468797616671617.

Turner, V. and Turner, E. (1978) *Image and Pilgrimage in Christian Culture*. Columbia University Press, New York.

Tutto Oggi (2018) 'Terremoto, nasce Comitato Rinascita Norcia: Le richieste' Tutto Oggi. Available at: https://tuttoggi.info/terremoto-nasce-comitato-rinascita-norcia-le-richieste/485517/ (accessed 29 July 2019).

UNESCO (2000) 24COM.XC.1 Assisi, The Basilica of San Francesco and other Franciscan Sites. *Report of the 24th Session of the World Heritage Committee*, Cairns, Australia, 27 November – 2 December.

12 Religion, the Environment, and Sacred Places: Lessons Learned and Future Directions

Daniel H. Olsen[1]* and Kiran A. Shinde[2]
[1]*Department of Geography, Brigham Young University, Provo, Utah, USA;* [2]*College of Arts, Social Sciences and Commerce, La Trobe University, Australia*

Introduction

The environment, with its various facets and dimensions, and sacred places and their intersection with religious tourism, have been the central focus of this book. In this final chapter, the authors reflect on some of the insights offered by the contributors on the interconnections between religious tourism and the environment that highlight the complexities surrounding the environment and sacred places. In outlining some of the limitations of this book, the authors also suggest avenues for future research in this area of inquiry.

Lessons Learned

As noted by the authors in Chapter 1, there are many angles from which the interrelationships between religion, tourism, and the environment can be approached. The key aspect of sacred sites is the physical or natural environment as conceptualized by Shinde and Olsen in Chapter 1: Fig. 1.2, p.10 and further developed by Shinde to propose a model for environmental analysis in a pilgrimage site in Chapter 3: Fig. 3.5, p. 55. As Olsen notes in Chapter 2, there are many religious, cultural, indigenous, and spiritual groups that consider certain aspects of the natural environment, such as mountains, volcanoes, caves, grottoes, bodies of water, trees, plants, and animals, sacred and worthy of being set apart and protected. These groups in many cases have developed deep theological and philosophical ideas related to the role of nature in the development of religious ritual and understandings of the place of humans within nature. Olsen outlines some of these religious views, focusing on how several religious groups have intertwined views of humanity with the natural environment. Olsen further notes that many global conservation groups and organizations also view these sacred natural sites as a type of world heritage and also important to preserve for biodiversity purposes. In Chapter 7, Farra-Haddad discusses some of these natural features in the context of interfaith pilgrimages in Lebanon, where trees, stones, and water are used by pilgrims to perform various religious rituals.

Nash, in Chapter 4, takes the relationship between religion and the environment deeper through an examination of the Vrindavan Ecological Concept (VEC) as developed by his teacher Shri Sevak Sharan (Sevakji). The VEC

*Corresponding author: dholsen@byu.edu

is comprised of seven principles of human ecology, which revolve around spirituality, nature, culture, tangible and intangible heritage, pilgrimage, human welfare, and ecology. These principles, when combined with seven types of consciousness as found in Hindu belief, lead a person to better understand the connection between humans, nature, and the divine. This then leads to the creation of what Nash calls a 'human sanctuary', which he defines as 'as a place-state where humans can live in a vibrant, non-static, and peaceful consonance with themselves as individuals and in an intimate and loving partnership with the greater human society and with nature' (p. 65).

In many instances, natural sacred sites are owned, maintained, and managed by religious groups, many of which, as a part of the sacralization process, modify natural environments by imposing sacral architecture onto these environments, such as temples, stupas, mosques, churches, and synagogues to create ideological and intentional sacred spaces. In other cases, sacred places are built with no reference to the natural environment. Regardless, these architectural modifications have in some cases been in place for centuries and are based on mythological legends, historical documentation, and miraculous accounts of divine irruption into the mundane world. The sacred nature of these natural, natural-human, and human-built sites inspires people to visit them for many reasons, including to worship, admire, gaze, be educated, and to relax and renew themselves. This process of visiting these sites because they are deemed sacred, as Bremer (2004) notes, reinforces their sacred or special nature. However, while visiting sacred sites leads to the development of a 'spirit of place', it also leads to the development of an infrastructure that is necessary to cater to or serve the needs of both local and visitors alike. This infrastructural development, along with the waste created by pilgrims and tourists, leads to the development of negative environmental outcomes.

In Chapter 10, Qurashi examines the negative environmental impacts of pilgrims who attend the *Hajj*. As Qurashi noted, there are several sources of waste at the *Hajj*, ranging from solid waste created by pilgrims and slaughterhouses to traffic congestion that causes severe air pollution. This has led to different types of environmental interventions by local and national government agencies, ranging from making waste less visible to increasing public transportation options to investment in better waste management systems and recycling plants. Olsen also addressed the issue of negative environmental outcomes from religious tourism and pilgrimage in Chapter 9 in the context of disease and other health-related issues that arise at mass religious gatherings. Olsen notes that the close proximity of, in some cases, millions of people can lead to the spread of communicable diseases both at these religious gatherings and within the generating regions from which these pilgrims come. This is particularly acute at the *Hajj*, where several diseases have been introduced to the pilgrim population and from there have spread to different world regions. While there have been several practices implemented at different scales to limit the spread of communicable diseases at mass religious gatherings, such as mandatory vaccinations, expanding and coordinating health facilities, and the use of modern technology to monitor outbreaks, dealing with the spread of disease is still an inexact science, particularly in the context of mass gatherings.

As an outcome of the development of negative environmental impacts due to pilgrimage and religious tourism, different kinds of social and spatial discourses regarding what constitute environmental problems, who is responsible for the presence and exacerbation of these problems, how to and address them, and, most importantly, who is responsible for rectifying these environmental issues, have evolved. This was demonstrated well by Shinde in Chapter 3 in his discussion of religious tourism development in six towns in India. Shinde notes that rises in pilgrim and religious tourism visitation have led to increased negative environmental impacts. However, while ritual acts by pilgrims add immensely to an already overburdened waste infrastructure, religious tourists, as well as tourism more generally, are blamed by both pilgrims and residents as the main culprits. This was also seen in Chapter 10, where Qurashi notes that pilgrims at the *Hajj* were indifferent when confronted with their unsustainable environmental practices, which views go against Islamic ideas of environmental stewardship and conservation.

Negative impacts of religious tourism on the natural and human-built environment, however, lead to negative impacts on host communities socially and economically. This is discussed in part in Chapter 5, where Nepal, Mu, and Lai discuss the perspectives of Sherpas in Nepal's Khumbu (Mt Everest) region in the context of religious tourism development in the Himalayas. As the authors note, the region has undergone drastic environmental, economic, and social changes due to the increased popularity of trekking tourism in the area. This has led to changes in the ways in which Sherpas view the sacred nature of the environment as well as leading to changes in Sherpas' lifestyle, economic patterns, education, marriage choices, and location of residence. While the younger generation of Sherpas feel that the environmental impacts of trekking have been overexaggerated, the older generation feel that the guardian spirits of the land are upset by these environmental and social changes, leading to tensions between the generations. And while 'tourism has not compromised Sherpa spiritual beliefs and cultural practices', it has rather 'reordered, reproduced, and realigned these beliefs and practices to adapt to a changing world' (p. 81).

Sacred places also require management and maintenance. In many cases, multiple stakeholders may wish to have a say in how a sacred site is managed, and questions may arise over ownership of these sites. As such, sacred site management and maintenance is inherently political. This is shown in Chapter 6, where Lafortune-Bernard, Suwal, Weise, and Coningham examine efforts to conserve the birthplace of Buddha in Lumbini, Nepal. While this site is a popular tourist attraction, there have been several issues relating to the conservation of the natural elements of the site. The development of a Lumbini Master Plan in 1978 has yet to be fully realized, in part because site managers have faced several challenges in implementing the plan due to stakeholders competing for the same resource base, a lack of monetary investment and community involvement, and limited coordination between different levels of government. These issues have led to the deterioration of the surrounding environment and greater difficulties in maintaining the sense of place of this famous religious site. In Chapter 8, Garcia-Fuentes takes a different perspective on the politics of sacred places, focusing on how Mt Monserrat in Spain was used by different political and religious groups to promote competing ideas of what constituted Catalonian identity. In particular, Garcia-Fuentes focuses on how Mt Monserrat was utilized by the famous architect Antoni Gaudí as the inspiration for the building of the Basílica de la Sagrada Família in Barcelona, and how the Basílica was subsequently used to solidify Catholic ideology. In sum, this case study reveals the entangled and sometimes hidden relations between religion, nature, and ideology.

In a slightly different vein, Farra-Haddad's chapter discusses how sacred places can be utilized to overcome politics and lead to deep and lasting interfaith dialogue. In Chapter 7, Farra-Haddad focuses on the ways in which pilgrims from various religious faiths share religious sites and use the same natural elements – trees, water, and stones – in shared religious rituals. She notes that these shared rituals lead to shared religious experiences, which lead to interfaith dialogue, peaceful living together, and broader regional goodwill. In Chapter 11, Di Giovine and Ascione look at the physical and spiritual after effects of earthquakes that shook the region of Central Italy in 2016 and how the affected communities dealt with destroyed religious architectural wonders and outmigration. One of the social outcomes of these natural disasters was the development of both a strong sense of community and different forms of solidarity with those that were affected the hardest to bring about different perceptions of sacredness. This sense of community and solidarity coalesced in many cases around a reimagination of sacred sites, where crumbled churches and basilicas became symbols of these communities, with religious tourism being reconceptualized to be more integrative of food, culture, and landscape, along with religion, to economically revitalize the region.

Limitations and Future Directions

As this volume serves as one of the first collection of essays focused on the environment and environmental aspects of religious tourism, in no way do the authors claim that the content, ideas, theories, and case studies in this text are comprehensive – at best this book is an entry point to the vast and diverse nature of this field of research, represented in part by the enclosed contributions. Indeed, like all texts, a cursory reading of this book will find limitations where more could have been discussed and expanded upon. For example, it is impossible to cover all geographic regions given the textual limitations of books, and there are many more places around the world where religious tourism leads to complicated relationships with different environments, whether natural or human-built (Terzidou *et al.*, 2008; Larson and Poudyal, 2012; Wong *et al.*, 2013; Sati, 2015; Alipour *et al.*, 2017). As such, additional case studies located in different world regions that would further enrich our understandings of the interconnections between religion, tourism, and the environment would be welcome and provide fertile ground for testing of some of the ideas presented in this book.

While mountains, grottoes, trees, stones, and water have been featured in this volume, rivers and larger water bodies, which is where most religious rituals are concentrated, are not represented here. Thankfully, there is a small literature relating to the environmental impacts of pilgrimages on rivers (Feldhaus, 1995; Alley, 2002; Haberman, 2006). Other natural formations and elements, such as caves and grottoes, are also severely impacted because of visitor flows (Kasim, 2011; Wang *et al.*, 2016; Paniandi *et al.*, 2018). Similarly, another productive research stream would be looking at how religious tourism drives urbanization processes and the resulting stresses on the environment that come with continued infrastructural development (Hussein *et al.*, 2016).

In this volume there has been little discussion, outside of Chapters 2, 4, and 7, looking at the effects of religious tourism from a theological perspective. Many of the chapters dealt with conceptual matters and draw from religious sites within Christianity, Hinduism, Islam, and Buddhism. But there are many indigenous sites that are oriented around natural landscapes, which is a glaring omission in this book (della Dora, 2012; Jimura, 2016; Suntikul and Dorji, 2016; Moufakkir and Selmi, 2018). Moreover, religious rituals, performances, and other devotional practices need to be investigated with greater detail, as was done in Farra-Haddad's chapter (Chapter 7), for a better understanding of their impacts on the environment (Huntsinger and Fernández-giménez, 2000; Kozihisseri, 2005; Chevrier, 2019; Moufahim and Lichrou, 2019).

Another area that should be investigated in greater detail is the issue of scale. While many of the case studies have looked at specific sacred sites, and although touched upon in Chapters 2 and 11, a greater examination of the effects of religious tourism on the environment at broader regional, national, and international scales would provide a greater understanding of the differential environmental impacts that affect larger areas. This might include a closer investigation of religious tourism along pilgrimage trails (Olsen and Trono, 2018), transboundary religious tourism, as occurs in the case of some Buddhist circuits (Choe and O'Regan, 2015), and how environmental resources, such as rivers and mountains, that are the focus of religious rituals are administered through cooperative institutional governance (Singh, 2005; Adler *et al.*, 2013; Sati, 2015). Similarly, it might be useful to understand how smaller local and regional sites compare with environmental issues found at more popular and iconic sites to understand the impact of overtourism or best practices relating to testing management techniques at smaller scales (McDuie-Ra and Chettri, 2018). More broadly, climate change and its impacts on sacred sites is an area that should be an emerging concern among scholars interested in the nexus of religious tourism and the environment. Mountains, valleys, rivers, lakes, deltas, and all forms of natural landscape that boast large numbers of sacred sites are vulnerable to climate change. There are instances where flooding has ravaged beautiful temples while shifting of waterways has led to the disappearance of sites of rituals, affecting the practice of pilgrimage and religious tourism (Haberman, 2006). Future studies need to incorporate climate change in any discussion of the management of sacred sites.

Another dimension that needs more careful attention is how temporary religious events, such as festivals and fairs, impact environmental mitigation efforts. Much of the early environmental research around pilgrimage was concerned with religious festivals and fairs as occasions where large crowds would pose the greatest threat for health, hygiene, and spread of diseases (Kamat, 2001; Maclean, 2003; Shinde, 2010). However, this line of research needs to be revisited given the changing nature and numbers of visitors at religious events and festivals (Raj *et al.*, 2015; Gupta and Basak, 2018).

As noted in this volume, most religions and faiths have clear ideas regarding showing reverence towards nature and encourage conservation of the natural environment. Many international and interfaith organizations seek to incorporate religious perspectives in their efforts to mitigate negative ecological footprints and encourage environmentally sustainable practices (COEJL, n.d.; Gottlieb, 2004). There are a few studies that have focused on how some of these organizations translate some of their environmental values such as deep-ecology, spiritual ecology, and eco-spirituality in their ritual practices (e.g. Tanner and Mitchell, 2002; Kiernan, 2015). However, translating these practices into a religious tourism context has not truly been explored outside of a few studies (Bagri *et al.*, 2009; Timothy, 2012; Liutikas, 2015; Kim *et al.*, 2019). Similarly, attention needs to be paid to governance structures, institutional arrangements, and site-level visitor management strategies in sacred places that are instituted to mitigate negative environmental impacts (Alley, 2000; Hu and Wall, 2005; Presti and Petrillo, 2010; Henderson, 2011; Shinde, 2011; Piva *et al.*, 2019).

There seems to be a correlation between the globalization of travel and the increasing environmental vulnerability of places. This is the case at religious centres, particularly those that experience increasing flows of pilgrims and religious tourists due in part to better transportation infrastructure. With tourism and pandemics like the coronavirus projected to continue to increase in the immediate future, religious sites will continue to face environmental pressures. The question is whether site managers can balance environmental protection and salvation.

References

Adler, C.E., McEvoy, D., Chhetri, P. and Kruk, E. (2013) The role of tourism in a changing climate for conservation and development. A problem-oriented study in the Kailash sacred landscape, Nepal. *Policy Sciences* 46(2), 161–178. DOI: 10.1007/s11077-012-9168-4.

Alipour, H., Olya, H.G.T. and Forouzan, I. (2017) Environmental impacts of mass religious tourism: from residents' perspectives. *Tourism Analysis* 22(2), 167–183. DOI: 10.3727/108354217X14888192562285.

Alley, K.D. (2000) Separate domains: Hinduism, politics, and environmental pollution. In: Chapple, C.K. and Tucker, M.E. (eds) *Hinduism and Ecology: The Intersection of Earth, Sky and Water*. Harvard University Press for the Center for the Study of World Religions, Cambridge, MA, pp. 355–381.

Alley, K.D. (2002) *On the Banks of the Ganga: When Waste Water Meets a Sacred River*. University of Michigan Press, Ann Arbor, Michigan.

Bagri, S., Gupta, B. and George, B. (2009) Environmental orientation and ecotourism awareness among pilgrims, adventure tourists, and leisure tourists. *Tourism* 57(1), 55–68.

Bremer, T.S. (2004) *Blessed With Tourists: The Borderlands of Religion and Tourism in San Antonio*. The University of North Carolina Press, Chapel Hill, North Carolina.

Chevrier, M.-H. (2019) Nocturnal ritual activities in tourist development of pilgrimage cities. *Journal of Policy Research in Tourism, Leisure and Events* 11(3), 436–454. DOI: 10.1080/19407963.2019.1585362.

Choe, J.Y. and O'Regan, M. (2015) Religious tourism experiences in South East Asia. In: Raj, R. and Griffin, K. (eds) *Religious Tourism and Pilgrimage Management: An International Perspective*, 2nd edn. CAB International, Wallingford, UK, pp. 191–204.

COEJL (n.d.) The coalition on the environment and Jewish life (COEJL). Available at: http://coejl.org (accessed 10 March 2020).

della Dora, V. (2012) Setting and Blurring boundaries: pilgrims, tourists, and landscape in Mount Athos and Meteora. *Annals of Tourism Research* 39(2), 951–974. DOI: 10.1016/j.annals.2011.11.013.

Feldhaus, A. (1995) *Water and Womanhood: Religious Meanings of Rivers in Maharashtra*. Oxford University Press, New York and Oxford.

Gottlieb, R.S. (ed.) (2004) *This Sacred Earth: Religion, Nature, Environment*. Routledge, New York.

Gupta, S. and Basak, B. (2018) Exploring pilgrim satisfaction on facilities for religious events: a case of Ratha Yatra at Puri. *Asia Pacific Journal of Tourism Research* 23(8), 765–779. DOI: 10.1080/10941665.2018.1486864.

Haberman, D.L. (2006) *River of Love in an Age of Pollution: The Yamuna River of Northern India*. University of California Press, Berkeley, California.

Henderson, J.C. (2011) Religious tourism and its management: the Hajj in Saudi Arabia. *International Journal of Tourism Research* 13(6), 541–552. DOI: 10.1002/jtr.825.

Hu, W. and Wall, G. (2005) Environmental management, environmental image and the competitive tourist attraction. *Journal of Sustainable Tourism* 13(6), 617–635. DOI: 10.1080/09669580508668584.

Huntsinger, L. and Fernandez-Gimenez, M. (2000) Spiritual pilgrims at Mount Shasta, California. *Geographical Review* 90(4), 536–558. DOI: 10.1111/j.1931-0846.2000.tb00353.x.

Hussein, A.O., Shahid, S., Basim, K.N. and Chelliapan, S. (2016) Modeling sewer flow in a pilgrimage city. *Journal of Environmental Engineering* 142(12), 05016005. DOI: 10.1061/(ASCE) EE.1943-7870.0001140.

Jimura, T. (2016) World heritage site management: a case study of sacred sites and pilgrimage routes in the Kii mountain range, Japan. *Journal of Heritage Tourism* 11(4), 382–394. DOI: 10.1080/1743873X.2016.1146287.

Kamat, M. (2001) The Palkhi as plague carrier: the Pandharpur fair and the sanitary fixation of the colonial state, British India, 1908–1916. In: Pati, B. and Harrison, M. (eds) *Health, Medicine and Empire*. Orient Longman, Hyderabad, India, pp. 299–316.

Kasim, A. (2011) Balancing tourism and religious experience: understanding devotees' perspectives on Thaipusam in Batu Caves, Selangor, Malaysia. *Journal of Hospitality Marketing & Management* 20(3-4), 441–456. DOI: 10.1080/19368623.2011.562437.

Kiernan, K. (2015) Landforms as sacred places: implications for geodiversity and geoheritage. *Geoheritage* 7(2), 177–193. DOI: 10.1007/s12371-014-0128-6.

Kim, H., Yilmaz, S. and Ahn, S. (2019) Motivational landscape and evolving identity of a route-based religious tourism space: a case of Camino de Santiago. *Sustainability* 11(13), 3547–47. DOI: 10.3390/su11133547.

Kozihisseri, D. (2005) Pilgrim's regress: Sabrimala rituals spell trouble for Kerala's Periyar reserve. Down To Earth. Available at: https://www.downtoearth.org.in/news/pilgrims-regress-10373 (accessed 10 March 2020).

Larson, L.R. and Poudyal, N.C. (2012) Developing sustainable tourism through adaptive resource management: a case study of Machu Picchu, Peru. *Journal of Sustainable Tourism* 20(7), 917–938. DOI: 10.1080/09669582.2012.667217.

Liutikas, D. (2015) Religious landscape and ecological ethics: pilgrimage to the Lithuanian Calvaries. *International Journal of Religious Tourism and Pilgrimage* 3(1), 12–24.

Maclean, K.K. (2003) Power and pilgrimage: the Kumbha Mela in Allahabad, 1765-1954. Ph.D. Thesis. La Trobe University.

McDuie-Ra, D. and Chettri, M. (2018) Himalayan boom town: rural-urban transformations in Namchi, Sikkim. *Development and Change* 49(6), 1471–1494. DOI: 10.1111/dech.12450.

Moufahim, M. and Lichrou, M. (2019) Pilgrimage, consumption and rituals: spiritual authenticity in a Shia Muslim pilgrimage. *Tourism Management* 70, 322–332. DOI: 10.1016/j.tourman.2018.08.023.

Moufakkir, O. and Selmi, N. (2018) Examining the spirituality of spiritual tourists: a Sahara desert experience. *Annals of Tourism Research* 70, 108–119. DOI: 10.1016/j.annals.2017.09.003.

Olsen, D.H. and Trono, A. (eds) (2018) *Religious Pilgrimage Routes and Trails: Sustainable Development and Management*. CABI, Wallingford, UK.

Paniandi, T.A., Albattat, A.R., Bijami, M., Alexander, A. and Balekrisnan, V. (2018) Marketing mix and destination image, case study: Batu Caves as a religious destination. *Almatourism-Journal of Tourism, Culture and Territorial Development* 9(17), 165–186.

Piva, E., Cerutti, S. and Raj, R. (2019) Managing the sacred: a governance perspective for religious tourism destinations. In: Griffiths, M. and Wiltshier, P. (eds) *Managing Religious Tourism*. CAB International, Wallingford, UK, pp. 10–21.

Presti, O.L. and Petrillo, C.S. (2010) Co-management of religious heritage: an Italian case-study. *Tourism* 58(3), 301–311.

Raj, R., Griffin, K. and Blackwell, R. (2015) Motivations for religious tourism, pilgrimage, festivals and events. In: Raj, R. and Griffin, K. (eds) *Religious Tourism and Pilgrimage Management: An International Perspective*, 2nd edn. CAB International, Wallingford, UK, pp. 103–117.

Sati, V.P. (2015) Pilgrimage tourism in mountain regions: Socio–economic and environmental implications in the Garhwal Himalaya. *South Asian Journal of Tourism and Heritage* 8(2), 164–182.
Shinde, K.A. (2010) Managing Hindu festivals in pilgrimage sites: emerging trends, opportunities, and challenges. *Event Management* 14(1), 53–67. DOI: 10.3727/152599510X12724735767598.
Shinde, K.A. (2011) 'This is religious environment': Environmental discourses and environmental behaviour in the Hindu pilgrimage site of Vrindavan. *Space and Culture* 14(4), 448–463.
Singh, S. (2005) Secular pilgrimages and sacred tourism in the Indian Himalayas. *GeoJournal* 64(3), 215–223. DOI: 10.1007/s10708-005-5649-8.
Suntikul, W. and Dorji, U. (2016) Local perspectives on the impact of tourism on religious festivals in Bhutan. *Asia Pacific Journal of Tourism Research* 21(7), 741–762. DOI: 10.1080/10941665.2015.1069365.
Tanner, R. and Mitchell, C. (2002) *Religion and the Environment*. Palgrave, New York.
Terzidou, M., Stylidis, D. and Szivas, E.M. (2008) Residents' perceptions of religious tourism and its socio-economic impacts on the Island of Tinos. *Tourism and Hospitality Planning & Development* 5(2), 113–129. DOI: 10.1080/14790530802252784.
Timothy, D. (2012) Religious views of the environment: Sanctification of nature and implications for tourism. In: Holden, A. and Fennell, D.A. (eds) *The Routledge Handbook of Tourism and the Environment*. Routledge, London and New York, pp. 53–64.
Wang, W., Chen, J.S. and Huang, K. (2016) Religious tourist motivation in Buddhist mountain: the case from China. *Asia Pacific Journal of Tourism Research* 21(1), 57–72. DOI: 10.1080/10941665.2015.1016443.
Wong, C.U.I., McIntosh, A. and Ryan, C. (2013) Buddhism and tourism: perceptions of the monastic community at Pu-Tuo-Shan, China. *Annals of Tourism Research* 40, 213–234.

Index

Note: Page numbers in **bold** type refer to **figures**
Page numbers in *italic* type refer to *tables*
Page numbers followed by 'n' refer to notes